T0335907

The Classical Orthogonal Polynomials

The Classical Orthogonal Polynomials

Brian George Spencer Doman

University of Liverpool, UK

World Scientific

NEW JERSEY · LONDON · SINGAPORE · BEIJING · SHANGHAI · HONG KONG · TAIPEI · CHENNAI · TOKYO

Published by

World Scientific Publishing Co. Pte. Ltd.

5 Toh Tuck Link, Singapore 596224

USA office: 27 Warren Street, Suite 401-402, Hackensack, NJ 07601

UK office: 57 Shelton Street, Covent Garden, London WC2H 9HE

Library of Congress Cataloging-in-Publication Data
Names: Doman, Brian George Spencer, 1936–
Title: The classical orthogonal polynomials / Brian George Spencer Doman,
    University of Liverpool, UK.
Description: New Jersey : World Scientific, 2015. | Includes bibliographical
    references and index.
Identifiers: LCCN 2015027954 | ISBN 9789814704038 (hardcover : alk. paper)
Subjects: LCSH: Orthogonal polynomials. | Polynomials.
Classification: LCC QA404.5 .D66 2015 | DDC 515/.55--dc23
LC record available at http://lccn.loc.gov/2015027954

British Library Cataloguing-in-Publication Data
A catalogue record for this book is available from the British Library.

Printed in Singapore

Contents

Preface

The Classical Orthogonal Polynomials have been studied extensively since the first set, the Legendre Polynomials, were discovered by Legendre in 1784. They frequently arise in the mathematical treatment of model problems in the Physical Sciences, often arising as solutions of ordinary differential equations subject to certain conditions imposed by the model. We shall not concern ourselves here with the physical applications. We shall be concentrating solely on their mathematical properties. This monograph derives a number of their basic properties together with some less well-known results.

The first chapter provides a survey of some general properties satisfied by any set of orthogonal polynomials. It starts by defining the inner product of two functions $f(x)$ and $g(x)$ as the integral of the product of these two functions multiplied by a non-negative weight function $w(x)$ over an interval (a, b), where a and b can be both finite, or one or both of infinite size. If this integral is zero, the functions $f(x)$ and $g(x)$ are said to be orthogonal. The functions we shall be considering are polynomials of arbitrary order.

The chapter continues by showing how for any given weight function we can use the orthogonality condition to produce a unique polynomial set by an iterative process and gives an example of this process. It shows that the orthogonality condition leads to a number of properties satisfied by any set of orthogonal polynomials. One of these is that members of any such set of polynomials satisfies a three-term recurrence relation. It also indicates that any set of polynomials satisfying such a recurrence relation forms an orthogonal set.

The first chapter then describes the particular choices of weight functions and domains which define the three classes of classical orthogonal polynomials. A number of additional properties of these classical orthog-

onal polynomials are then deduced. In particular it is shown that each of
the polynomials satisfies a second order differential equation.

Each subsequent chapter focusses on a particular orthogonal polynomial
set starting from the viewpoint of its differential equation. It shows that
solutions of this differential equation which satisfy certain conditions are
polynomials and that these polynomials form an orthogonal set. It then
describes in detail a number of the properties outlined in chapter 1 together
with further interesting properties.

A number of the polynomials have the Gamma Function $\Gamma(z)$ as part of
their definition. The Gamma Function is defined in the General Appendix
and the properties used in this monograph derived. The Beta Function
$B(a, b)$ and the Hypergeometric Function $_2F_1(a, b; c; z)$ are also defined and
the properties which are used in the earlier chapters described.

This monograph is an expanded version of a series of projects devised
for undergraduate mathematicians at Liverpool University.

Chapter 1

Definitions and General Properties

1.1   Introduction

The classical orthogonal polynomials arise in a number of practical situations and models, often as solutions to differential equations arising from boundary value problems. We shall be collecting together and examining a number of their properties without detailed reference to their applications.

In this chapter we look at the definition for a set of orthogonal polynomials and describe a process for their generation and a number of their main properties. We then specialise to the classical orthogonal polynomials and deduce an additional number of general properties and show in particular that they all satisfy a second order differential equation. It is this approach from a differential equation which often arises in practical applications.

In later chapters we examine the properties of each of the individual polynomial sets from a different viewpoint, starting from the differential equation.

1.2   Definition of Orthogonality

The scalar or inner product of two functions $f(x)$ and $g(x)$ is defined by the integral

$$\int_a^b w(x)f(x)g(x)dx, \tag{1.2.1}$$

where $w(x) \geq 0$, for $a \leq x \leq b$.

This is a generalisation of the idea of a scalar product of two finite dimensional vectors to an infinite dimensional "function space". If this scalar

1

product is zero, we say that the functions $f(x)$ and $g(x)$ are orthogonal. Here we shall be looking at functions $R_n(x)$ which are polynomials of order n.

If these nth order polynomials $R_n(x)$ satisfy the orthogonality relation

$$\int_a^b w(x) R_n(x) R_m(x) dx = 0 \qquad \text{for} \qquad n \neq m, \qquad (1.2.2)$$

where $w(x)$ is a weight function which is non-negative in the interval (a, b) and is such that the integral is well-defined for all finite order polynomials $R_n(x)$, these polynomials form a set of orthogonal polynomials. It is clear that

$$\int_a^b w(x) [R_n(x)]^2 dx = h_n \geq 0 \qquad (1.2.3)$$

since the integrand is everywhere ≥ 0 for $a < x < b$.

1.3   Gram-Schmidt Orthogonalisation Procedure

For a given weight function $w(x)$, this is an inductive procedure to generate a set of orthogonal polynomials starting from the zeroth order polynomial $R_0(x) = 1$. The procedure works by using the orthogonality condition to determine the coefficients of the powers of x in the polynomial $R_{n+1}(x)$ using all of the previously determined $R_m(x), 0 \leq m \leq n$.

The procedure consists of the following steps:

Set $R_0(x) = 1$.

Set $R_1(x) = x + a_{1,0}$. The constant $a_{1,0}$ is determined by the orthogonality condition:

$$\int_a^b w(x)(x + a_{1,0}) dx = 0 = \int_a^b x\, w(x) dx + a_{1,0} \int_a^b w(x) dx. \qquad (1.3.1)$$

If the integral $\int_a^b x\, w(x) dx = 0$, then $a_{1,0} = 0$.

Set $R_2(x) = x^2 + a_{2,1} x + a_{2,0}$. The constants $a_{2,0}$ and $a_{2,1}$ are determined by the conditions that $R_2(x)$ is orthogonal to $R_1(x)$ and $R_0(x)$ that is:

$$\int_a^b w(x)\left(x^2 + a_{2,1} x + a_{2,0}\right) dx = 0 \qquad (1.3.2)$$

and

$$\int_a^b w(x)\left(a_{1,0} + x\right)\left(x^2 + a_{2,1} x + a_{2,0}\right) dx = 0, \qquad (1.3.3)$$

or equivalently, Eq. (1.3.2) and

$$\int_a^b w(x)x\big(x^2 + a_{2,1}x + a_{2,0}\big)dx = 0. \qquad (1.3.3.1)$$

If we denote $I_n = \int_a^b w(x)x^n dx$, then the coefficients $a_{2,1}$ and $a_{2,0}$ are determined by the matrix equation

$$\begin{pmatrix} I_0 & I_1 \\ I_1 & I_2 \end{pmatrix} \begin{pmatrix} a_{2,0} \\ a_{2,1} \end{pmatrix} = - \begin{pmatrix} I_2 \\ I_3 \end{pmatrix}$$

which determines $a_{2,0}$ and $a_{2,1}$ provided that the determinant

$$\Delta_2 = \begin{vmatrix} I_0 & I_1 \\ I_1 & I_2 \end{vmatrix} \neq 0. \qquad (1.3.4)$$

For $R_3(x) = x^3 + a_{3,2}x^2 + a_{3,1}x + a_{3,0}$, we have three constants to determine. These are determined by using the orthogonality of $R_3(x)$ and $R_0(x)$, $R_1(x)$ and $R_2(x)$, or equivalently, the orthogonality of $R_3(x)$ and x^n for $n = 0$, 1 and 2. The constants $a_{3,0}$, $a_{3,1}$ and $a_{3,2}$ can then be determined uniquely provided that

$$\Delta_3 = \begin{vmatrix} I_0 & I_1 & I_2 \\ I_1 & I_2 & I_3 \\ I_2 & I_3 & I_4 \end{vmatrix} \neq 0. \qquad (1.3.5)$$

Continuing this process, we define $R_n(x) = x^n + \sum_{m=0}^{n-1} a_{n,m}x^m$ and use the orthogonality relations with $R_0(x)$, $R_1(x)$,..., $R_{n-1}(x)$, or equivalently setting $R_n(x)$ orthogonal to x^m for $0 \leq m < n$. That is

$$\int_a^b w(x)R_n(x)x^m dx = 0 \qquad \text{for all} \qquad m < n. \qquad (1.3.6)$$

These equations can be written in matrix form

$$\begin{pmatrix} I_0 & I_1 & I_2 & \dots & I_{n-1} \\ I_1 & I_2 & I_3 & \dots & I_n \\ \dots & \dots & \dots & \dots & \dots \\ I_{n-1} & I_n & I_{n+1} & \dots & I_{2n-2} \end{pmatrix} \begin{pmatrix} a_{n,0} \\ a_{n,1} \\ \dots \\ a_{n,n-1} \end{pmatrix} = - \begin{pmatrix} I_n \\ I_{n+1} \\ \dots \\ I_{2n-1} \end{pmatrix}.$$

This will have a unique solution provided that the $n-1$ by $n-1$ determinant Δ_{n-1} of the integrals I_m, $0 \leq m \leq 2n - 2$, is not zero. The coefficients $a_{n,m}$ can be evaluated using Cramer's rule.

If we include

$$\int_a^b w(x)x^n R_n(x)dx = \int_a^b w(x)[R_n(x)]^2 dx = h_n,$$

these equations can be put into the matrix form

$$
\begin{pmatrix}
I_0 & I_1 & I_2 & \cdots & I_n \\
I_1 & I_2 & I_3 & \cdots & I_{n+1} \\
\cdots & \cdots & \cdots & \cdots & \cdots \\
I_n & I_{n+1} & I_{n+2} & \cdots & I_{2n}
\end{pmatrix}
\begin{pmatrix}
a_{n,0} \\
a_{n,1} \\
\cdots \\
1
\end{pmatrix}
=
\begin{pmatrix}
0 \\
0 \\
\cdots \\
h_n
\end{pmatrix}.
\tag{1.3.7}
$$

From these equations we see that $h_n = \Delta_n/\Delta_{n-1}$. In fact we see that taking $\Delta_0 = I_0$,

$$
\Delta_n = h_n h_{n-1} h_{n-2}...h_1 \Delta_0
$$

and therefore $\Delta_n > 0$ for all n.

We can represent the transfomation from the functions 1, x, x^2, ..., x^N to $R_0(x)$, $R_1(x)$, ..., $R_N(x)$ by the matrix transformation

$$
\begin{pmatrix}
R_0(x) \\
R_1(x) \\
R_2(x) \\
...... \\
R_N(x)
\end{pmatrix}
=
\begin{pmatrix}
1 & 0 & 0 & .... & 0 \\
a_{1,0} & 1 & 0 & .... & 0 \\
a_{2,0} & a_{2,1} & 1 & .... & 0 \\
.... & ..... & ..... & .... & 0 \\
a_{N,0} & a_{N,1} & a_{N,2} & .... & 1
\end{pmatrix}
\begin{pmatrix}
1 \\
x \\
x^2 \\
... \\
x^N
\end{pmatrix}.
\tag{1.3.8}
$$

The determinant of this transformation matrix A is 1. This tranformation matrix A with the coefficients $a_{i,j}$ is not singular and can therefore be inverted. The inverse of this matrix is also a lower triangular matrix with leading diagonal elements 1. It follows that any power of x, say x^M, can be represented as a linear combination of the polynomials $R_m(x)$ with $0 \leq m \leq M$. This means that it also follows that any Mth order polynomial $Q_M(x)$ can be represented as a linear combination of the $R_m(x)$ with $0 \leq m \leq M$ and therefore that

$$
\int_a^b w(x) R_N(x) Q_M(x) dx = 0 \qquad \text{for all} \qquad M < N.
\tag{1.3.9}
$$

As an example of the orthogonalisation procedure, we construct the polynomials defined between $-\pi/2$ and $\pi/2$ with a weighting factor of $\cos x$. The first few of these polynomials are:

$C_0 = 1$.

$C_1 = x$.

$C_2 = x^2 + 2 - \pi^2/4 = x^2 - 0.4674011$.

$C_3 = x^3 - (\pi^4 - 48\pi^2 + 384)x/(4\pi^2 - 32) = x^3 - 1.02536x$.

$C_4 = x^4 - 1.610684x^2 + 0.27358$.

$C_5 = x^5 - 2.207395x^3 + 0.883659x$.

$C_6 = x^6 - 2.823442x^4 + 1.860313x^2 - 0.1695036$.

These polynomials do not appear to have any practical application.

The graphs of C_1 to C_6 are plotted below

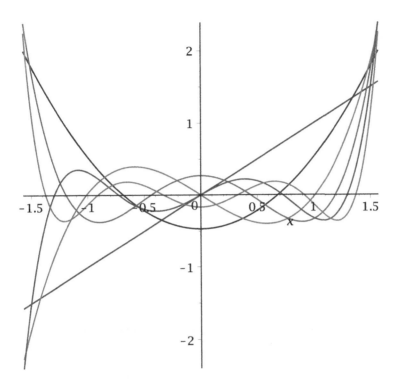

In the following sections we describe some general properties satisfied by all orthogonal polynomial sets.

1.4   The Nth Order Polynomial $R_N(x)$ has N Distinct Real Zeros in the Interval (a, b)

The orthogonality condition between $R_N(x)$ and $R_0(x)$ for $N > 0$,

$$\int_a^b w(x)R_N(x)dx = 0 \qquad (1.4.1)$$

implies that since $R_N(x)$ is not zero everywhere, it must be positive for some values of x in (a, b) and negative for some other values of x in (a, b). Since $R_N(x)$ is continuous, there will be a point x_1 in the interval (a, b)

where $R_N(x)$ changes sign. Thus $R_N(x)$, $N > 0$ has a zero at $x = x_1$ in this interval.

If $N > 1$, $R_N(x)$ is orthogonal to all polynomials of order 1. In particular $\int_a^b w(x)(x - x_1)R_N(x)dx = 0$, $N > 1$.

This means that $(x - x_1)R_N(x)$ changes sign at some point x_2 in the interval (a, b). This point x_2 is different from x_1 since $R_N(x)$ changes sign there. Therefore $R_N(x)$, $N > 1$ has another zero at $x = x_2$, $a < x_2 < b$.

If $N > 2$, $R_N(x)$ is orthogonal to all polynomials of order 2. In particular $\int_a^b w(x)(x - x_1)(x - x_2)R_N(x)dx = 0$.

The integrand will change sign at a point x_3 in the interval (a, b), and x_3 must be different from x_1 and x_2.

This process can be continued to show that $R_N(x)$ has N distinct real zeros in the interval (a, b).

From this result it follows that if $x_1...x_n$ are the zeros of $R_n(x)$,

$$R_n(x) = \prod_{i=1}^{n}(x - x_i). \qquad (1.4.2)$$

1.5   Gauss Quadrature Formula

This formula is of importance in the numerical analysis for the approximate evaluation of integrals. Our interest however is more theoretical. We follow the formulation in Chihara.

If $x_{n,k}$ are the zeros of the polynomial $R_n(x)$, then for any polynomial $Q_m(x)$, $m \leq 2n - 1$

$$\int_a^b w(x)Q_m(x)dx = \sum_{k=1}^{n} A_{n,k}Q_m(x_{n,k}). \qquad (1.5.1)$$

The numbers $A_{n,k}$ are all real and positive and satisfy the condition

$$\sum_{k=1}^{n} A_{n,k} = \int_a^b w(x)dx = I_0 \qquad (1.5.2)$$

which can be obtained by substituting $Q_m(x) = 1$ in Eq. (1.5.1).

Proof   Let $\{s_1, ..., s_n\}$ be a set of n distinct real numbers $a < s_i < b$, then

$$F(x) = \prod_{i=1}^{n}(x - s_i)$$

is a polynomial of degree n and $F(x)/(x - s_j)$ is a polynomial of degree $n - 1$. Consider

$$\lim_{x \to s_j} \frac{F(x)}{x - s_j} = F'(s_j) \neq 0$$

since s_i are distinct, and

$$l_j(x) = \frac{F(x)}{(x - s_j)F'(s_j)}$$

is a polynomial of degree $n - 1$ with the property

$$l_j(s_k) = \delta_{k,j}$$

where $\delta_{k,j}$ is the Kronecker delta which equals 1 if $k = j$ and zero otherwise.

The $(n - 1)$ th order Lagrange Interpolating Polynomial approximation to the N th order polynomial $Q_N(x)$ is then

$$S(x) = \sum_{i=1}^{n} l_i(x) Q_N(s_i). \tag{1.5.3}$$

If $N < n$, $Q_N(x)$ is exactly equal to $S(x)$ above and its integral is given by Eq. (1.5.5) below. If $N \geq n$, $Q_N(x) - S(x)$ is a polynomial with n zeros at s_i and can therefore be written as the product of an $(N - n)$ th order polynomial $T_{N-n}(x)$ multiplied by $\prod_{i=1}^{n}(x - s_i)$. The integral of $Q_N(x)$ becomes

$$\int_a^b w(x) Q_N(x) dx = \int_a^b w(x) T_{N-n}(x) \prod_{i=1}^{n}(x - s_i) dx + \sum_{i=1}^{n} A_{n,i} Q_N(s_i) \tag{1.5.4}$$

where $A_{n,i} = \int_a^b w(x) l_i(x) dx$. We now set $\prod_{i=1}^{n}(x - s_i) = R_n(x)$ where $R_n(x)$ is the n th order orthogonal polynomial for the weight function $w(x)$. This means that $s_i = x_{n,i}$, the i th zero of $R_n(x)$. The first integral is then zero for any polynomial $T_{N-n}(x)$ of order less than n. This means that

$$\int_a^b w(x) Q_N(x) dx = \sum_{i=1}^{n} A_{n,i} Q_N(x_{n,i}) \tag{1.5.5}$$

is exact for all polynomials $Q_N(x)$ of order $2n - 1$ or less.

Noting that $l_i(x)$ is a polynomial of order $n - 1$ we can write

$$0 < \int_a^b w(x) l_k^2(x) dx = \sum_{i=1}^{n} A_{n,i} l_k^2(x_{n,i}) = A_{n,k} \tag{1.5.6}$$

which shows that $A_{n,k}$ is positive.

1.6 Recurrence Relation

In this section we show that the polynomials $R_N(x)$ satisfy a recurrence relation of the form

$$R_{N+1}(x) = (\beta_N x - \alpha_N)R_N(x) - \gamma_N R_{N-1}(x). \qquad (1.6.1)$$

If the polynomials $R_N(x)$ are monic, that is if the coefficient of x^n is 1, as would be the case if they were generated by the Gramm-Schmidt process, the coefficient $\beta_N = 1$. This follows from equating coefficients of x^{N+1} in the above equation. It then follows that $R_{N+1}(x) - xR_N(x)$ is a polynomial of order N and therefore is a linear combination of $R_m(x)$ for $0 \le m \le N$.

$$R_{N+1}(x) - xR_N(x) = \sum_{n=0}^{N} b_n R_n(x). \qquad (1.6.2)$$

On the left side of this equation, the coefficient of x^N is $a_{N+1,N} - a_{N,N-1}$, where $a_{n,m}$ is the coefficient of x^m in the polynomial $R_n(x)$, and that on the right is just b_N. Therefore in Eq. (1.6.1), $\alpha_N = -b_N = a_{N,N-1} - a_{N+1,N} = k'_N - k'_{N+1}$, where k'_N is the coefficient of x^{n-1} in $R_N(x)$.

Alternatively, if we multiply the sum by $R_N(x)$ and integrate from a to b, we find

$$-\int_a^b xR_N^2(x)w(x)dx = b_N \int_a^b R_N^2(x)w(x)dx$$

and therefore

$$\alpha_N = -b_N = \int_a^b xw(x)R_N^2(x)dx \bigg/ \int_a^b w(x)R_N^2(x)dx. \qquad (1.6.3)$$

Thus $R_{N+1}(x) - (x - \alpha_N)R_N(x)$ becomes a polynomial of order $N - 1$ and can be represented as a linear combination of $R_n(x)$, $0 \le n \le N - 1$, that is

$$R_{N+1}(x) - (x - \alpha_N)R_N(x) = \sum_{n=0}^{N-1} b_n R_n(x). \qquad (1.6.4)$$

We can work out the coefficients b_n by multiplying Eq. (1.6.4) by $w(x)R_m(x)$ with $0 \le m < N$, and integrating from a to b. Using the orthogonality of the $R_n(x)$, we see that the right-hand side reduces to

$$b_m \int_a^b w(x)[R_m(x)]^2 dx = b_m h_m,$$

while the left-hand side becomes

$$\int_a^b w(x)R_{N+1}(x)R_m(x)dx + \alpha_N \int_a^b w(x)R_N(x)R_m(x)dx$$

$$- \int_a^b w(x)xR_N(x)R_m(x)dx. \tag{1.6.5}$$

The first and second integrals are zero for all m in $0 \le m \le N - 1$ since $R_{N+1}(x)$ and $R_N(x)$ are orthogonal to $R_m(x)$. In the third integral we note that $xR_m(x)$ is a polynomial of order $m + 1$. This means that this integral will be zero unless $m + 1 \ge N$. It follows that $b_m = 0$ unless $m = N - 1$. It follows that Eq. (1.6.1) is true with $\gamma_N = -b_{N-1}$.

If we consider the third integral in Eq. (1.6.5) for $m = N - 1$, we note that $xR_{N-1}(x) = x^N + \sum_{n=0}^{N-2} a_{N-1,n}x^{n+1}$. Since x^{n+1} is orthogonal to $R_N(x)$ for $0 \le n \le N - 2$, none of the terms in the sum contribute to this integral. The integral

$$\int_a^b w(x)x^N R_N(x)dx = \int_a^b w(x)[R_N(x)]^2 dx = h_N,$$

where $h_N = \int_a^b w(x)[R_N(x)]^2 dx$. Therefore $\gamma_N = -b_{N-1} = h_N/h_{N-1}$. This leaves us with the recurrence relation:

$$R_{N+1}(x) = (x + k'_{N+1} - k'_N)R_N(x) - \frac{h_N}{h_{N-1}}R_{N-1}(x). \tag{1.6.6}$$

We note at this point that γ_N in the recurrence relation

$$R_{N+1}(x) = (x - \alpha_N)R_N(x) - \gamma_N R_{N-1}(x) \tag{1.6.7}$$

is positive and α_N is real and

$$k'_n = \sum_{m=1}^{n-1} \alpha_m. \tag{1.6.8}$$

We note that for the monic polynomials $R_N(x)$,

$$h_N = \gamma_N h_{N-1} = \gamma_N \gamma_{N-1}...\gamma_1 h_0. \tag{1.6.9}$$

We now show that it follows from this recurrence relation that the polynomial $R_n(x)$ has n distinct zeros and that the zeros of $R_n(x)$ lie between the zeros of $R_{n+1}(x)$. Let $x_{n,i}$ be the ith zero of $R_n(x)$, where $x_{n,n} < x_{n,n-1} < ... < x_{n,1}$. We will examine monic polynomials but multiplicating by a factor will not affect these results. Consider first $x_{n,1}$ and suppose $x_{n-1,1} < x_{n,1}$, then

$$R_{n+1}(x_{n,1}) = -\gamma_n R_{n-1}(x_{n,1}) \quad < \quad 0, \tag{1.6.10}$$

since γ_n is positive and $R_{n-1}(x_{n,1})$ is also positive since $R_{n-1}(x)$ is positive for $x > x_{n-1,1}$. For large values of x, $R_{n+1}(x)$ is positive and so $R_{n+1}(x)$ has a zero at $x_{n+1,1} > x_{n,1}$.

Let us now suppose that the zeros of $R_{n-1}(x)$ interlace the zeros of $R_n(x)$, that is $x_{n,i-1} < x_{n-1,i} < x_{n,i}$ so that the sign of $R_{n-1}(x_{n,i})$ is $(-1)^{i+1}$. Then the sign of

$$R_{n+1}(x_{n,i}) = -\gamma_n R_{n-1}(x_{n,i})$$

is $(-1)^i$. Thus $R_{n+1}(x_{n,1})$ is negative, $R_{n+1}(x_{n,2})$ is positive and so on. This means that $R_{n+1}(x)$ has a zero between $x_{n,i}$ and $x_{n,i-1}$. There are $n-1$ of these zeros. For the last zero, the sign of $R_{n+1}(x_{n,n})$ is $(-1)^n$. If n is odd, $R_{n+1}(x)$ becomes positive for large negative values of x and must therefore have a zero $x_{n+1,n+1} < x_{n,n}$. Similarly if n is even, $R_{n+1}(x)$ becomes negative for large negative values of x and so again, $R_{n+1}(x)$ has a zero $x_{n+1,n+1} < x_{n,n}$. It is easy to see that the zeros of $R_2(x)$ interlace the zeros of $R_3(x)$. It therefore follows by induction that the zeros of $R_n(x)$ interlace the zeros of R_{n+1}. Also we note that $x_{n+1,1} > x_{n,1}$ so that $x_{n,1}$ form an increasing sequence and as $x_{n-1,n-1} < x_{n,n}$, $x_{n,n}$ form a decreasing sequence.

1.7 Favards Theorem

This theorem was announced by Favard in 1935 but was discovered independently by Shohat and Natanson.

In this section we will denote by $\mathcal{L}$ the linear moment functional which for a polynomial $R_n(x)$ can be represented by the integral

$$\mathcal{L}[R_n(x)] = \int_a^b w(x) R_n(x)(x) dx, \qquad (1.7.1)$$

provided that the integral converges.

The formal definition of $\mathcal{L}$ consists of the specification of a set of real numbers $\{\mu_n\}_{n=0}^{\infty}$ such that

$$\mathcal{L}[x^n] = I_n, \qquad (1.7.2)$$

together with the linearity property

$$\mathcal{L}[aR_n(x) + bR_m(x)] = a\mathcal{L}[R_n(x)] + b\mathcal{L}[R_m(x)]$$

for any polynomials $R_n(x)$ and $R_m(x)$ and constants a and b.

Favard's theorem is essentially the converse of the result of the previous section. That is given a set of arbitrary complex numbers c_n and γ_n, and the polynomials defined by the recurrence formula

$$R_0(x) = 1 \qquad\qquad R_{-1} = 0$$

$$R_{n+1}(x) = (x - c_n)R_n(x) - \gamma_n R_{n-1}(x). \qquad (1.7.3)$$

There is a unique moment funtional $\mathcal{L}$ for which these polynomials form an orthogonal set. That is $\mathcal{L}[R_n(x)R_m(x)] = 0$ if $n \neq m$.

Proof The proof proceeds by using the recurrence relation to define $\mathcal{L}[x^n]$ for progressively higher powers of x using the condition $\mathcal{L}[R_n(x)] = 0$.

Starting with

$$\mathcal{L}[R_0(x)] = I_0,$$

$$\mathcal{L}[R_1(x)] = \mathcal{L}[x - c_0] = \mathcal{L}[x] - c_0 I_0 = I_1 - c_0 I_0 = 0.$$

This defines $I_1 = c_0 I_0$.

$$\mathcal{L}[R_2(x)] = 0 = \mathcal{L}[(x - c_1)(x - c_0) - \gamma_1] = I_2 - (c_1 + c_0)I_1 - \gamma_1 I_0.$$

This defines I_2. The higher order moments $I_n = \mathcal{L}[x^n]$ are then defined successively using the recurrence relation.

If we write Eq. (1.7.3) in the form

$$xR_n(x) = R_{n+1}(x) + c_n R_n(x) + \gamma_n R_{n-1}(x), \qquad (1.7.4)$$

we see that

$$\mathcal{L}[xR_n(x)] = 0, \qquad\qquad n \geq 2 \qquad (1.7.5)$$

since the moment functional of each of the terms on the right is zero. We note that

$$\mathcal{L}[xR_1(x)] = \gamma_1 \mathcal{L}[R_0(x)] = \gamma_1 I_0. \qquad (1.7.6)$$

If we multiply Eq. (1.7.4) by x and take note of the above equation, we find

$$\mathcal{L}[x^2 R_n(x)] = 0 \qquad\qquad n \geq 3 \qquad (1.7.7)$$

and

$$\mathcal{L}[x^2 R_2(x)] = \gamma_2 \mathcal{L}[xR_1(x)] = \gamma_2 \gamma_1 \mathcal{L}[R_0(x)] = \gamma_2 \gamma_1 I_0. \qquad (1.7.8)$$

Proceeding in this way, we see that

$$\mathcal{L}[x^m R_n(x)] = 0 \qquad\qquad n > m \qquad (1.7.9)$$

and so

$$\mathcal{L}[Q_m(x)R_n(x)] = 0 \qquad n > m \qquad (1.7.10)$$

for any polynomial $Q_m(x)$. This shows that $R_n(x)$ form an orthogonal set. Also

$$\mathcal{L}[x^n R_n(x)] = \mathcal{L}[(R_n(x))^2] = \gamma_n...\gamma_2\gamma_1 I_0 \qquad (1.7.11)$$

since $R_n(x)$ is a monic polynomial.

We see that the polynomials $R_n(x)$ will all be real if all of c_n and γ_n are real and that the moment functional $\mathcal{L}[(R_n(x))^2]$ will be positive definite if $\gamma_n > 0$ for all n.

To develop the representation of the moment functional $\mathcal{L}$ we firstly use the Gaussian Quadrature formula from section 5.

$$\mathcal{L}[Q_M(x)] = \int_a^b w(x)Q_M(x)dx = \sum_{k=1}^N A_{N,k}Q_M(x_{N,k})$$

where $x_{N,k}$ are the zeros of the N th order polynomial $R_N(x)$ of the set of orthogonal polynomials generated by the recurrence relation. This is exact for all $M < 2N$. Thus

$$\mathcal{L}[x^m] = I_m = \sum_{k=1}^N A_{N,k}x_{N,k}^m \qquad m < 2N. \qquad (1.7.12)$$

As the N zeros $x_{N,k}$ are all distinct, this relation could in principle be used to determine $A_{N,k}$. If we now denote $\Delta x_{N,k} = x_{N,k} - x_{N,k+1}$, then

$$\mathcal{L}[Q_M(x)] = \sum_{k=1}^{N-1} \frac{A_{N,k}}{\Delta x_{N,k}} Q_M(x_{N,k})\Delta x_{N,k} + A_{N,N}Q_N(x_{N,N}). \qquad (1.7.13)$$

If we now set $a = \lim_{N\to\infty} x_{N,N}$ and $b = \lim_{N\to\infty} x_{N,1}$, this would indicate that in the limit $N \to \infty$, Eq. (1.7.13) would become

$$\mathcal{L}[Q_M(x)] = \int_a^b w(x)Q_M(x)dx, \qquad (1.7.14)$$

where the weight function $w(x) = \lim_{N\to\infty} A_{N,k}/\Delta x_{N,k}$.

The arguments in the preceding paragraph are merely suggestive. For a rigorous discussion for the implementation of the moment functional as an integral, the reader is referred to the book by Chihara.

1.8   The Christoffel-Darboux Formula

If we multiply Eq. (1.6.1) with $\beta_N = 1$ by $R_N(y)$,

$$R_{N+1}(x)R_N(y) = (x - \alpha_N)R_N(x)R_N(y) - \frac{h_N}{h_{N-1}}R_{N-1}(x)R_N(y). \quad (1.8.1)$$

If we now interchange x and y and subtract from Eq. (1.8.1),

$$R_{N+1}(x)R_N(y) - R_{N+1}(y)R_N(x)$$

$$= (x - y)R_N(x)R_N(y) + \frac{h_N}{h_{N-1}}[R_N(x)R_{N-1}(y) - R_N(y)R_{N-1}(x)]. \quad (1.8.2)$$

If we now iterate this formula, we obtain the Christoffel-Darboux formula

$$R_{N+1}(x)R_N(y) - R_{N+1}(y)R_N(x) = (x-y)h_N \sum_{n=0}^{N} R_n(x)R_n(y)/h_n. \quad (1.8.3)$$

If we divide through by $(x - y)$ and take the limit $y \to x$ we find

$$R_N(x)R'_{N+1}(x) - R'_N(x)R_{N+1}(x) = h_N \sum_{n=0}^{N}[R_n(x)]^2/h_n. \quad (1.8.4)$$

1.9   Interlacing of Zeros

We can use the Christoffel-Darboux formula to show that there is one and only one zero of $R_n(x)$ between every zero of $R_{n+1}(x)$. We note first of all that the right-hand side of Eq. (1.8.4) is always positive. Therefore if x_1 is a zero of $R_{n+1}(x)$, $R_n(x_1)R'_{n+1}(x_1)$ is positive. If x_1 and x_2 are successive zeros of $R_{n+1}(x)$, then $R'_{n+1}(x_1)$ and $R'_{n+1}(x_2)$ will have opposite signs since the zeros are both simple zeros. Therefore $R_n(x_1)$ and $R_n(x_2)$ will also have opposite signs. Since $R_n(x)$ is a continuous function of x, it must pass through zero somewhere between x_1 and x_2. Now $R_{n+1}(x)$ has $n + 1$ separate zeros and between each of these is a zero of $R_n(x)$, and so there can only be one zero between successive zeros of $R_{n+1}(x)$.

A similar argument can be used to show that there is a zero of $R_{n+1}(x)$ between any two successive zeros of $R_n(x)$. We note that if x_k is a zero of $R_n(x)$, $R'_n(x_k)R_{n+1}(x_k)$ is negative. Using the above argument, we see that $R_{n+1}(x)$ changes sign between any two successive zeros of $R_n(x)$.

1.10   Minimum Property

Of all of the polynomials above with coefficient of $x^N = 1$, the polynomials $R_N(x)$ above have the smallest difference from zero in the mean. That is

$$\int_a^b w(x)[R_N(x)]^2 dx$$

is smaller than for any other polynomial with coefficient of $x^N = 1$. For if $S_N(x)$ is another such polynomial, $R_N(x) - S_N(x)$ is a polynomial of order $N - 1$ which can be expanded in terms of $R_n(x)$.

$$S_N(x) - R_N(x) = \sum_{n=0}^{N-1} a_n R_n(x).$$

Then

$$\int_a^b w(x)[S_N(x)]^2 dx - \int_a^b w(x)[R_N(x)]^2 dx = \int_a^b w(x)[S_N(x) - R_N(x)]^2 dx$$

$$= \sum_{n=0}^{N-1} a_n^2 \int_a^b [R_n(x)]^2 dx \ge 0. \qquad (1.10.1)$$

The first equality follows from the fact that

$$\int_a^b w(x) R_N(x)[S_N(x) - R_N(x)] dx = 0.$$

This proves the minimum property.

1.11   Approximation of Functions

In this section we examine the approximation of a function $\phi(x)$ by a sum of the polynomials $R_n(x)$, $S_N(x) = \sum_{n=0}^N a_n R_n(x)$. We show how to calculate the coefficients in this sum to minimise the weighted mean square error

$$\int_a^b w(x)[\phi(x) - S_N(x)]^2 dx = \int_a^b w(x)[\phi(x)]^2 dx$$

$$-2\sum_{n=0}^N a_n \int_a^b w(x)\phi(x)R_n(x)dx + \sum_{n=0}^N a_n^2 \int_a^b w(x)[R_n(x)]^2 dx.$$

We have used the orthogonality condition in the last term above. Minimisation of the above expression with respect to variations with respect to the a_n gives

$$a_n \int_a^b w(x)[R_n(x)]^2 dx = \int_a^b w(x)\phi(x)R_n(x)dx. \qquad (1.11.1)$$

The weighted mean square error is then

$$\int_a^b w(x)[\phi(x)]^2 dx - \sum_{n=0}^N a_n^2 \int_a^b w(x)[R_n(x)]^2 dx. \tag{1.11.2}$$

1.12   Definitions of Some Parameters

k_N is the coefficient of x^N in the polynomial $R_N(x)$. For the polynomials defined by the Gramm Schmidt process above, $k_N = 1$ but the classical orthogonal polynomials have different values for k_N.

k_N' is the coefficient of x^{N-1} in the polynomial $R_N(x)$.

$h_N = \int_a^b w(x)[R_N(x)]^2 dx$.

The next sections describe some general properties satisfied by all of the Classical Orthogonal Polynomials. More detailed descriptions and specific applications are described in the chapters on the individual functions.

1.13   Fundamental Intervals and Weight Functions of the Classical Orthogonal Polynomials

The interval (a, b) can be transformed to one of $(-1, 1)$, $(0, \infty)$ or $(-\infty, \infty)$ as follows:

(i) a and b both finite: let $x = [a + b + (b - a)t]/2$ so that when $t = -1$, $x = a$ and when $t = 1$, $x = b$.

(ii) a finite, $b = \infty$: in this case we transform to the interval $(0, \infty)$ by writing $x = t - a$.

(iii) $a = -\infty$ and $b = \infty$.

The classical orthogonal polynomials can then be separated into three distinct groups.

The first group, the Hermite polynomials $H_n(x)$, are defined for the domain $-\infty < x < \infty$ with a weight function $w(x) = e^{-x^2}$.

The second group, the Associated Laguerre polynomials $L_n^{(\alpha)}(x)$, are defined for $0 \le x < \infty$ with a weight function $w(x) = x^\alpha e^{-x}$, $\alpha > -1$.

The third group, the Jacobi polynomials $P_n^{(\alpha,\beta)}(x)$ and all of its special cases, are defined for the domain $-1 \le x \le 1$ with a weight function $w(x) = (1 - x)^\alpha (1 + x)^\beta$, $\alpha, \beta > -1$.

The domains and weight functions for the Associated Laguerre polynomials and the Hermite polynomials can be derived from those for the Jacobi polynomials by a process of shifting, scaling and then taking a limit. For the Associated Laguerre polynomials, we shift the domain to $0 \leq x \leq b$ with weight function $x^\alpha (1 - x/b)^b$ and then taking the limit $b \to \infty$. For the Hermite polynomials we have the domain $-a \leq x \leq a$ with weight function $[(1 - x/a)(1 + x/a)]^{a^2}$ and take the limit $a \to \infty$.

1.14   Recurrence Relations

We note here the recurrence relations for the Hermite polynomials $H_N(x)$, the Associated Laguerre polynomials $L_N^{(\alpha)}(x)$ and for the Jacobi polynomials $P_N^{(\alpha,\beta)}(x)$ (see also Hochstrasser Urs W). These relations can be derived from the general recurrence relation Eq. (1.6.1) using the appropriate coefficient of x^N or more easily directly following the methods in the following chapters on the functions themselves.

Hermite polynomials $H_N(x)$

$$H_{N+1}(x) = 2xH_N(x) - 2NH_{N-1}(x). \tag{1.14.1}$$

Associated Laguerre polynomials $L_N^{(\alpha)}(x)$

$$(N+1)L_{N+1}^\alpha(x) = (2N + \alpha + 1 - x)L_N^\alpha(x) - (N + \alpha)L_{N-1}^\alpha(x). \tag{1.14.2}$$

Jacobi polynomials $P_N^{(\alpha,\beta)}(x)$

$$2(N+1)(N + \alpha + \beta + 1)(2N + \alpha + \beta)P_{N+1}^{(\alpha,\beta)}(x)$$

$$= \left[(2N + \alpha + \beta + 1)(\alpha^2 - \beta^2) \right.$$

$$\left. + (2N + \alpha + \beta)(2N + \alpha + \beta + 1)(2N + \alpha + \beta + 2)x \right] P_N^{(\alpha,\beta)}(x)$$

$$-2(N + \alpha)(N + \beta)(2N + \alpha + \beta + 2)P_{N-1}^{(\alpha,\beta)}. \tag{1.14.3}$$

1.15   Differential Relation

In this section we show that for each of the Classical Orthogonal polynomials a multiple of the derivative, $R_N'(x)$ can be expressed as a linear combination of $R_N(x)$ and $R_{N-1}(x)$ with coefficients dependent on x. Firstly we note that $dR_N(x)/dx$ is a polynomial of order $N-1$ and the coefficient of x^{N-1} is N times the coefficient of x^N in the polynomial $R_N(x)$.

At this point we have to consider the different types of interval separately.

Hermite polynomials $H_N(x)$

The interval is $(-\infty, \infty)$ and the weight e^{-x^2}.

$H'_N(x)$ is a polynomial of order $N-1$ which can be represented as a linear combination of Hermite polynomials up to order $N-1$. Consider

$$\frac{d}{dx} H_N(x) = \sum_{n=0}^{N-1} b_n H_n(x).$$

We multiply this by $\exp(-x^2)H_m(x)$ and integrate from $-\infty$ to ∞. On using the orthogonality relations, we see that the right-hand side becomes $h_m b_m$. The left-hand side becomes

$$\int_{-\infty}^{\infty} e^{-x^2} \frac{dH_N(x)}{dx} H_m(x)dx.$$

On integrating by parts, this becomes

$$\left[e^{-x^2} H_N(x) H_m(x) \right]_{-\infty}^{\infty} - \int_{-\infty}^{\infty} e^{-x^2} \left[e^{x^2} \frac{d}{dx} \left(e^{-x^2} H_m(x) \right) \right] H_N(x) dx.$$

The integrated part tends to zero at both end points. The expression inside the square bracket is a polynomial of order $m+1$ and therefore integral is zero if $m < N-1$. This means that $H'_N(x)$ is a multiple of $H_{N-1}(x)$. That is

$$\frac{d}{dx} H_N(x) = \mu H_{N-1}(x). \tag{1.15.1}$$

Now $H_N(x) = 2^N x^N + O(x^{N-1})$. Therefore $H'_N(x) = 2^N N x^{N-1} + O(x^{N-2})$. Equating powers of x^{N-1} on both sides leads to $\mu = 2N$.

Associated Laguerre polynomials $L_N^{(\alpha)}(x)$

The interval is $(0, \infty)$ and the weight function $x^\alpha e^{-x}$, where $\alpha > -1$. $x \, dL_N^{(\alpha)}(x)/dx$ is a polynomial of order N whose coefficient of x^N is N times the coefficient of x^N in $L_N^{(\alpha)}(x)$ and therefore

$$x\frac{d}{dx} L_N^{(\alpha)}(x) - N L_N^{(\alpha)}(x) = \sum_{n=0}^{N-1} b_n L_n^{(\alpha)}(x). \tag{1.15.2}$$

We now multiply both sides by $x^\alpha e^{-x} L_m^{(\alpha)}(x)$ where $m < N$, and integrate from 0 to ∞. The right-hand side becomes $h_m b_m$. We integrate the integral of the first term on the left by parts to get

$$\int_0^\infty x^{\alpha+1} e^{-x} \frac{dL_N^{(\alpha)}(x)}{dx} L_m^{(\alpha)}(x) dx$$

$$= \left[x^{\alpha+1} e^{-x} L_N^{(\alpha)}(x) L_m^{(\alpha)}(x) \right]_0^\infty - \int_0^\infty L_N^{(\alpha)}(x) \frac{d}{dx} \left(x^{\alpha+1} e^{-x} L_m^{(\alpha)}(x) \right).$$

The integrated part vanishes at both limits of the integration. The other term can be written

$$- \int_0^\infty x^\alpha e^{-x} L_N^{(\alpha)}(x) \left[x^{-\alpha} e^x \frac{d}{dx} \left(x^{\alpha+1} e^{-x} L_m^{(\alpha)}(x) \right) \right] dx.$$

It is not difficult to see that the expression inside the square brackets is a polynomial of order $m+1$ and that this integral is therefore zero if $m+1 < N$.

The integral of the other term on the left of Eq. (1.15.2), for $m < N$, $N \int_0^\infty x^\alpha e^{-x} L_N^{(\alpha)}(x) L_m^{(\alpha)}(x) dx$, is zero because of the orthogonality of the Associated Laguerre polynomials. Thus we have the relation

$$x \frac{d}{dx} L_N^{(\alpha)}(x) = N L_N^{(\alpha)}(x) + \mu L_{N-1}^{(\alpha)}(x). \qquad (1.15.3)$$

The value of $\mu = b_{N-1} = -k_N'/k_{N-1}$ can be obtained by equating coefficients of x^{N-1} in (1.15.3).

Jacobi polynomials $P_N^{(\alpha,\beta)}(x)$

The interval is $(-1,1)$ and the weight function $w(x) = (1-x)^\alpha (1+x)^\beta$, where $\alpha, \beta > -1$.

A large number of special cases for particular values of α and β are known by different names. We shall study these in later sections.

For the expression

$$(1-x^2) \frac{d}{dx} P_N^{(\alpha,\beta)}(x) + N x P_N^{(\alpha,\beta)}(x),$$

we note that the coefficient of x^{N+1} vanishes and therefore this expression is a polynomial of order N and so can be represented as a sum of Jacobi polynomials in the form

$$(1-x^2) \frac{d}{dx} P_N^{(\alpha,\beta)}(x) + N x P_N^{(\alpha,\beta)}(x) = \sum_{n=0}^N b_n P_n^{(\alpha,\beta)}(x). \qquad (1.15.4)$$

We now multiply both sides by $w(x) P_m^{(\alpha,\beta)}(x)$ and integrate from -1 to 1. From the orthogonality relation we see that the right-hand side becomes $h_m b_m$. Firstly let us examine the integral of the first term of the left-hand side

$$\int_{-1}^1 w(x)(1-x^2) P_m^{(\alpha,\beta)}(x) \frac{d}{dx} P_N^{(\alpha,\beta)}(x) dx.$$

On integration by parts, this becomes

$$\left[(1-x^2)w(x)P_N^{(\alpha,\beta)}(x)P_m^{(\alpha,\beta)}(x)\right]_{-1}^{1}$$

$$-\int_{-1}^{1} P_N^{(\alpha,\beta)}(x)dx\frac{d}{dx}\left(w(x)(1-x^2)P_m^{(\alpha,\beta)}(x)\right).$$

The integrated part vanishes at both end points and so we are left with

$$-\int_{-1}^{1} w(x)P_N^{(\alpha,\beta)}(x)dx\left[\frac{1}{w(x)}\frac{d}{dx}\left(w(x)(1-x^2)P_m^{(\alpha,\beta)}(x)\right)\right].$$

It is not difficult to see that the expression inside the square brackets is a polynomial of order $m+1$ and that the integral is therefore zero if $m < N-1$.

For the integral of the second term on the left-hand side of Eq. (1.15.4),

$$N\int_{-1}^{1} w(x)P_N^{(\alpha,\beta)}(x)xP_m^{(\alpha,\beta)}(x)dx,$$

we note that $xP_m^{(\alpha,\beta)}(x)$ is a polynomial of order $m+1$ and so this integral is also zero if $m < N-1$. This means that all of the expansion coefficients in Eq. (1.15.4) are zero except b_N and b_{N-1} so that

$$(1-x^2)\frac{d}{dx}P_N^{(\alpha,\beta)}(x) = -NxP_N^{(\alpha,\beta)}(x) + b_N P_N^{(\alpha,\beta)}(x) + b_{N-1}P_{N-1}^{(\alpha,\beta)}(x).$$
$$(1.15.5)$$

Equating the coefficients of x^N on both sides gives $b_N = k'_N/k_N$. b_{N-1} can be found by equating the coefficients of x^{N-1} on both sides of the equation. Another way to evaluate b_N and b_{N-1} is by putting $x = \pm 1$ in Eq. (1.15.5). This is detailed in chapter 11 as is an alternative way to evaluate b_N and b_{N-1}.

The recurrence relations in the previous section can be used to express the polynomial $R_{N-1}(x)$ in terms of $R_N(x)$ and $R_{N+1}(x)$. This can then be used to express the derivatives in equations (1.15.1), (1.15.3) and (1.15.5) in terms of $R_N(x)$ and $R_{N+1}(x)$.

1.16 Step Up and Step Down Operators

Equations (1.15.1), (1.15.3) and (1.15.5) can be used to express the orthogonal polynomial $R_{N-1}(x)$ in terms of $R_N(x)$ and its derivative.

$$R_{N-1}(x) = f_N^-(x)R_N(x) + g_N^-(x)R'_N(x). \qquad (1.16.1)$$

For Hermite polynomials, f_N^- is zero and $g_N^- = 1/(2N)$.

For the associated Laguerre polynomials, $f_N^- = -N/\mu$ and $g_N^- = x/\mu$, where $\mu = b_{N-1} = -k_N'/k_{N-1}$.

For the Jacobi polynomials, $f_N^- = (Nx - b_N)/b_{N-1}$ and $g_N^- = (1-x^2)/b_{N-1}$.

We can regard

$$S_N^- = \left(f_N^-(x) + g_N^-(x) \frac{d}{dx} \right) \tag{1.16.2}$$

as an operator acting on $R_N(x)$ to produce $R_{N-1}(x)$, that is a step down operator.

We can use the general form of the recurrence relation (1.6.1) to derive an expression for R_{N+1} in terms of R_N and its derivative in the form

$$R_{N+1}(x) = f_N^+(x)R_N(x) + g_N^+(x)R_N'(x). \tag{1.16.3}$$

This gives rise to the step up operator

$$S_N^+ = \left(f_N^+(x) + g_N^+(x) \frac{d}{dx} \right). \tag{1.16.4}$$

We note that the coefficients g_N^- and g_N^+ do not change sign throughout the domains which define the polynomials.

1.17   Interlacing of Zeros

Using the step down and step up operators of the previous section leads to an alternative (simpler) derivation of the interlacing zeros property which we noted earlier.

If x_1 and x_2 are successive zeros of $R_N(x)$, the derivatives of $R_N(x)$ at these zeros will have opposite signs, not being zero since all of the zeros are simple. The values of $R_{N-1}(x_1) = g_N^-(x_1)R_N'(x_1)$ and $R_{N-1}(x_2) = g_N^-(x_2)R_N'(x_2)$ will also have different signs and so $R_{N-1}(x)$ must be equal to zero for some value of x, $x_1 < x < x_2$.

The same argument can be used to show that there will be a zero of $R_{N+1}(x)$ between x_1 and x_2.

1.18   Differential Equation

Each of the Classical Orthogonal polynomials satisfies a second order linear ordinary differential equation. We can obtain the differential equation for each of the polynomials by applying the step down and step up operators in succession.

Specifically, we note that
$$R_N(x) = S_{N-1}^+ S_N^- R_N(x) = S_{N+1}^- S_N^+ R_N(x). \tag{1.18.1}$$
Since $S_N^\pm$ is a first order differential operator, Eq. (1.18.1) is a second order differential equation.

Bochner considered second order differential equations of the form
$$a_2(x)\frac{d^2y}{dx^2} + a_1(x)\frac{dy}{dx} + a_0(x)y + \lambda_n y = 0, \tag{1.18.2}$$
where $a_0(x)$, $a_1(x)$ and $a_2(x)$ are polynomials. He showed that if this equation is to have solutions which are polynomials of the order n, $a_0(x)$ must be a constant which by redefining λ_0 can be taken to be zero. To have a solution $y = x - c$, $a_1(x) + \lambda_1(x - c) = 0$ and so $a_1(x)$ must be a first order polynomial. Finally, for a solution $y = x^2 + bx + c$ we see that $a_2(x)$ must be a polynomial of order 2 or less.

In the following chapters we will examine the properties of the individual functions in detail.

References

Bochner S, Uber Sturm-Liouvillesche Polynomsysteme, Mat. Zeit., 29 (1929), 730-736.

Chihara T S, An Introduction to Orthogonal Polynomials, Gordon and Breach, 1978.

Courant R and Hilbert D, Methods of Mathematical Physics Vol 1, Interscience Publishers, 1953.

Dennery P and Krzywicki A, Mathematics for Physicists, Harper and Rowe, 1967.

Erdelyi A, Higher Transcendental Functions Vol 2, McGraw-Hill, 1953.

Favard J, Sur les Polynomes de Tchebysheff, C. R. Acad. Sci. Paris, 200 (1935), 2052-2053.

Hochstrasser Urs W, Orthogonal Polynomials, Chapter 22, Handbook of Mathematical Functions, Eds. Abramowitz M and Stegun I A, Dover, 1970.

Koornwinder T H, Wong R, Koekoek R and Swarttouw R F, Chapter 18, NIST Handbook of Mathematical Functions, Eds. Olver W J, Lozier D W, Boisvert R F and Clark C W, NIST and Cambridge University Press, 2009.

Natanson I, Konstruktive Functionentheorie, Akademie-Verlag, Berlin 1955.

Shohat J, Theorie Generale des Polynomes Orthogonaux de Tchebyshef, Memoires des Sciences Mathematiques, 66 (1934) 1-68.

Chapter 2

Hermite Polynomials

2.1 Introduction

Hermite polynomials are defined in two different ways by different classes of users. For the physicists, Hermite polynomials denoted by $H_n(x)$ arise from following the orthogonalisation process for polynomials defined over a domain $(-\infty, \infty)$ with a weight function $\exp(-x^2)$. The probabilists define the Hermite polynomials, commonly denoted by $He_n(x)$, over the domain $(-\infty, \infty)$ with the weight function $\exp(-x^2/2)$. The two functions are related by $H_n(x) = 2^{n/2} He_n(\sqrt{2}\,x)$. $He_n(x)$ are thus monic polynomials. [See Koornwinder et al., Hochstrasser or Wikipedia]. In this chapter we will use the physicists' definitions but it is a simple matter to transform the various properties for the probabilists' functions.

We can think of $H_n(x)$ as arising from the domain (a, b) with weight function $w(x) = (x - a)^\alpha (b - x)^\beta$. If we let $b = -a$, and $\alpha = \beta = a^2$, $w(x)$ becomes $(1 - x^2/a^2)^{a^2}$. If we take the limit $a \to \infty$, $w(x)$ becomes $\exp(-x^2)$.

The Hermite polynomials are related to the Laguerre polynomials by

$$H_{2n}(x) = (-4)^n n! L_n^{(-1/2)}(x^2) = (-1)^n \frac{(2n)!}{n!}\, {}_1F_1(-n; 1/2; x^2) \qquad (2.1.1)$$

and

$$H_{2n+1}(x) = 2(-4)^n n! x L_n^{(1/2)}(x^2) = (-1)^n \frac{(2n+1)!}{n!} 2x\, {}_1F_1(-n; 3/2; x^2)$$
$$(2.1.2)$$

If we follow the Gram-Schmidt orthogonalisation procedure described in the first chapter and then multiply the polynomial $R_n(x)$ by 2^n so that in the resulting polynomial the coefficient of x^n is 2^n, we obtain the Hermite polynomials $H_n(x)$.

The first few Hermite polynomials are

$H_0(x) = 1.$

$H_1(x) = 2x.$

$H_2(x) = 4x^2 - 2.$

$H_3(x) = 8x^3 - 12x.$

$H_4(x) = 16x^4 - 48x^2 + 12.$

$H_5(x) = 32x^5 - 160x^3 + 120x.$

$H_6(x) = 64x^6 - 480x^4 + 720x^2 - 120.$

$H_7(x) = 128x^7 - 1344x^5 + 3360x^3 - 1680x.$

$H_8(x) = 256x^8 - 3584x^6 + 13440x^4 - 13440x^2 + 1680.$

$H_9(x) = 512x^9 - 9216x^7 + 48384x^5 - 80640x^3 + 30240x.$

$H_{10}(x) = 1024x^{10} - 23040x^8 + 161280x^6 - 403200x^4 + 302400x^2 - 30240.$

Graphs of $\exp(-x^2/2)H_n(x)$ for $n = 1$ to $n = 5$.

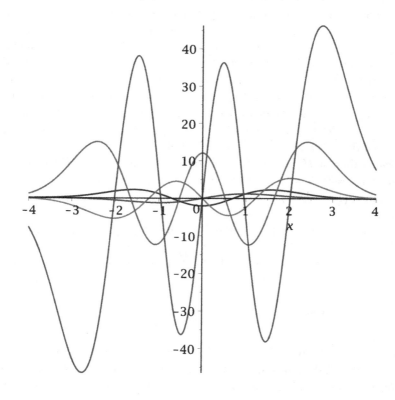

2.2 Differential Equation

In many applications Hermite polynomials arise as polynomial solutions of a second order ordinary differential equation. We shall examine this differential equation and show that these polynomial solutions are orthogonal with weight factor $\exp(-x^2)$ and so must be multiples of the Hermite polynomials.

The differential equation satisfied by Hermite polynomials is

$$\frac{d^2y}{dx^2} - 2x\frac{dy}{dx} + \lambda y = 0. \tag{2.2.1}$$

The point $x = 0$ is an ordinary point. This means that the solution can be written in the form of a power series in x. There are no singular points for finite x and so the series solution will converge for all finite values of x. If we write the solution as the power series $y = \sum_{n=0}^{\infty} a_n x^n$ and substitute into the differential equation we get

$$\sum_{n=0}^{\infty} n(n-1)a_n x^{n-2} - 2\sum_{n=0}^{\infty} a_n x^n + \lambda \sum_{n=0}^{\infty} a_n x^n = 0.$$

Equating the coefficient of x^n to zero leads to the recurrence relation

$$\frac{a_{n+2}}{a_n} = \frac{2n - \lambda}{(n+2)(n+1)}. \tag{2.2.2}$$

Since this ratio tends to zero as $n \to \infty$, both of these solutions converge for all finite values of x. We thus have two infinite series solutions. One containing only odd powers of x and one with only even powers.

If $\lambda = 2m$ for some integer m, then one of the infinite series solutions terminates and becomes an mth order polynomial. This polynomial will consist only of even powers of x if m is an even number and only odd powers if m is odd.

2.3 Orthogonality

We can write the differential equation (2.2.1) for the polynomial solution $R_m(x)$ in Sturm Liouville form.

$$\frac{d}{dx}\left(e^{-x^2}\frac{dR_m(x)}{dx}\right) + 2me^{-x^2}R_m(x) = 0. \tag{2.3.1}$$

If we multiply this by $R_n(x)$, where $n \neq m$ and integrate from $-\infty$ to ∞ we obtain

$$\int_{-\infty}^{\infty} R_n(x)\frac{d}{dx}\left(e^{-x^2}\frac{dR_m(x)}{dx}\right) = -2m\int_{-\infty}^{\infty} e^{-x^2}R_n(x)R_m(x)dx.$$

If we integrate the integral on the left-hand side by parts and note that as e^{-x^2} times any polynomial tends to zero as $x \to \pm\infty$, the integrated part vanishes. This leads to the equation

$$\int_{-\infty}^{\infty} e^{-x^2} R_n'(x) R_m'(x) dx = 2m \int_{-\infty}^{\infty} e^{-x^2} R_n(x) R_m(x) dx.$$

If we now follow the procedure above with n and m interchanged, we end up with the same integral on the left-hand side and the same integral on the right but multiplied by $2n$ instead of $2m$. If we take the difference of these two equations we get

$$0 = 2(m - n) \int_{-\infty}^{\infty} e^{-x^2} R_n(x) R_m(x) dx.$$

It follows that since $m \neq n$

$$\int_{-\infty}^{\infty} e^{-x^2} R_n(x) R_m(x) dx = 0. \tag{2.3.2}$$

This is the orthogonality relation satisfied by the Hermite polynomials. Therefore if we multiply $R_n(x)$ by a constant so that the coefficient of x^n is 2^n, we will obtain the Hermite polynomial $H_n(x)$. Since the series consists of only even powers of x or only odd powers, the coefficient $k_n' = 0$.

2.4   Derivative Property

The pth derivative of $H_n(x)$ is a polynomial of order $n - p$. We now show that it is a multiple of $H_{n-p}(x)$. If we differentiate the differential equation for the Hermite polynomial $H_n(x)$ p times we find with $y = d^p H_n(x)/dx^p$

$$\frac{d^2 y}{dx^2} - 2x \frac{dy}{dx} + (2n - 2p)y = 0. \tag{2.4.1}$$

The polynomial solution of this equation is $y = aH_{n-p}(x)$ which is therefore a multiple of the pth derivative of $H_n(x)$. We can find the constant by considering the highest powers of x. The highest power term in $H_n(x)$ is $(2x)^n$. Differentiated p times, it becomes $2^n x^{n-p} n!/(n-p)!$. Therefore the pth derivative of $H_n(x)$ is

$$\frac{d^p}{dx^p} H_n(x) = \frac{2^p n!}{(n-p)!} H_{n-p}(x). \tag{2.4.2}$$

2.5   Rodrigues Formula

We firstly note that

$$\frac{d^n}{dx^n}e^{-x^2} \to 0 \qquad \text{when} \qquad x \to \pm\infty$$

since it is e^{-x^2} times a polynomial. It then follows that for $n \neq 0$

$$\int_{-\infty}^{\infty} \frac{d^n}{dx^n}e^{-x^2}dx = 0.$$

If we integrate by parts we see that

$$\int_{-\infty}^{\infty} x\frac{d^n}{dx^n}e^{-x^2}dx = 0. \qquad n > 1,$$

and integrating by parts m times

$$\int_{-\infty}^{\infty} x^m\frac{d^n}{dx^n}e^{-x^2}dx = 0 \qquad \text{provided that} \qquad m < n. \qquad (2.5.1)$$

In other words, the nth order polynomial $Q_n(x) = e^{x^2}d^ne^{-x^2}/dx^n$, with weighting factor e^{-x^2} is orthogonal to x^m for all values of $n > m$. Thus $Q_n(x)$ is orthogonal to any polynomial of order $m < n$ and in particular $Q_n(x)$ is orthogonal to $Q_m(x)$ for all values of $n > m$. These polynomials must therefore be multiples of the Hermite polynomials. It is easy to see that the coefficient of x^n in $Q_n(x)$ is $(-2)^n$ and so $H_n(x) = (-1)^nQ_n(x)$.
    Thus

$$H_n(x) = (-1)^ne^{x^2}\frac{d^n}{dx^n}e^{-x^2}. \qquad (2.5.2)$$

2.6   Explicit Expression

An explicit power series expansion can be obtained by repeated use of the recurrence relation (2.2.2). We find

$$H_n(x) = n!\sum_{m=0}^{N}\frac{(-1)^m}{m!(n-2m)!}(2x)^{n-2m}, \qquad (2.6.1)$$

where $N = [n/2]$, the largest integer $\leq n/2$.
    Alternatively,

$$H_{2m}(x) = (-1)^m\sum_{p=0}^{m}\frac{(2m)!}{(m-p)!(2p)!}(-4x^2)^p \qquad (2.6.2)$$

and

$$H_{2m+1}(x) = (-1)^m 2x \sum_{p=0}^{m} \frac{(2m+1)!}{(m-p)!(2p+1)!}(-4x^2)^p. \tag{2.6.3}$$

The coefficient of x^n in $H_n(x)$, k_n is

$$k_n = 2^n. \tag{2.6.4}$$

We can express x^n in terms of Hermite polynomials

$$x^n = \sum_{k=0}^{\infty} a_k H_k(x).$$

Multiplying both sides by $\exp(-x^2)H_j(x)$ and integrating from $-\infty$ to ∞ gives

$$\int_{-\infty}^{\infty} e^{-x^2} x^n H_j(x) dx = \sum_{k=0}^{\infty} a_k \int_{-\infty}^{\infty} e^{-x^2} H_j(x) H_k(x) dx = h_j a_j,$$

since $H_n(x)$ are an orthogonal set. Therefore

$$a_j h_j = \int_{-\infty}^{\infty} x^n (-1)^j \frac{d^j}{dx^j} e^{-x^2} dx = \frac{n!}{(n-j)!} \int_{-\infty}^{\infty} e^{-x^2} x^{n-j} dx.$$

If $n - j$ is odd, this integral will be zero. Let $n - j = 2m$ then

$$a_{n-2m} h_{n-2m} = \frac{n!}{(2m)!} \int_{-\infty}^{\infty} x^{2m} e^{-x^2} dx = \frac{n!}{(2m)!} \frac{(2m)!\sqrt{\pi}}{2^{2m} m!}.$$

Then

$$x^n = \sum_{m=0}^{[n/2]} \frac{n!}{2^n m!(n-2m)!} H_{n-2m}(x), \tag{2.6.5}$$

where $[n/2]$ is the largest integer $p \leq n/2$.

2.7   Generating Function

We start with the identity

$$\frac{\partial^2}{\partial x^2} e^{2xt-t^2} - 2x \frac{\partial}{\partial x} e^{2xt-t^2} = -2t \frac{\partial}{\partial t} e^{2xt-t^2}. \tag{2.7.1}$$

Let us define $\phi_n(x)$ by

$$\sum_{n=0}^{\infty} \frac{t^n}{n!} \phi_n(x) = e^{2xt-t^2} \qquad |t| < 1. \tag{2.7.2}$$

The functions $\phi_n(x)$ are nth order polynomials in x. We can see this by differentiating (2.7.2) by t n times and putting $t = 0$ or by simply writing the right-hand side of (2.7.2) as an expansion in powers of t.

If we substitute the left-hand side of (2.7.2) into (2.7.1) we obtain

$$\sum_{n=0}^{\infty} \frac{t^n}{n!} \left\{ \frac{d^2\phi_n(x)}{dx^2} - 2x\frac{d\phi_n(x)}{dx} \right\} = -2t \sum_{n=0}^{\infty} \frac{\phi_n(x)}{n!} \frac{d}{dt} t^n = - \sum_{n=0}^{\infty} 2n \frac{t^n}{n!} \phi_n(x).$$
(2.7.3)

If we now equate the coefficients of powers of t on both sides of the equation, we find that

$$\frac{d^2\phi_n(x)}{dx^2} - 2x\frac{d\phi_n(x)}{dx} = -2n\phi_n(x). \tag{2.7.4}$$

In other words $\phi_n(x)$ satisfies Hermite's equation and is therefore some multiple of the Hermite polynomial $H_n(x)$.

We can calculate $\phi_n(x)$ from Eq. (2.7.2) as follows

$$\phi_n(x) = \lim_{t \to 0} \frac{\partial^n}{\partial t^n} e^{2xt-t^2} = e^{x^2} \lim_{t \to 0} \frac{\partial^n}{\partial t^n} e^{-(x-t)^2}$$

$$= (-1)^n e^{x^2} \lim_{t \to 0} \frac{\partial^n}{\partial x^n} e^{-(x-t)^2} = (-1)^n e^{x^2} \frac{d^n}{dx^n} e^{-x^2},$$

which is the Rodrigues formula. Therefore $\phi_n(x) = H_n(x)$, the Hermite polynomial.

The generating function for the Hermite polynomials is thus

$$\sum_{n=0}^{\infty} \frac{t^n}{n!} H_n(x) = e^{2xt-t^2} \qquad |t| < 1. \tag{2.7.5}$$

The generating function can be used to prove a number of identities. It can be used to show that the functions $\phi_n(x)$ defined in (2.7.2) are orthogonal and must therefore be multiples of the Hermite polynomials.

We can use the generating function to evaluate $h_n = \int_{-\infty}^{\infty} e^{-x^2} H_n^2(x)dx$. Consider the product of expansion (2.7.2) for $\phi_n(x)$ in powers of t and the expansion for $\phi_m(x)$ in powers of s. Multiply this by e^{-x^2} and integrate from $-\infty$ to ∞. This produces

$$\sum_{n,m=0}^{\infty} \int_{-\infty}^{\infty} e^{-x^2} \frac{t^n}{n!} \phi_n(x) \frac{s^m}{m!} \phi_m(x)dx = \int_{-\infty}^{\infty} e^{-x^2} e^{2tx-t^2} e^{2sx-s^2} dx.$$

On completing the square in the exponent, this becomes

$$\int_{-\infty}^{\infty} e^{-(x-t-s)^2} e^{2st} dx = \sqrt{\pi} e^{2st}.$$

If we expand e^{2st} as a power series, the general term is $2^n s^n t^n / n!$, there are no terms of the form $s^m t^n$ with $n \neq m$. The coefficient of $t^n s^m$ on the left-hand side is $\int_{-\infty}^{\infty} e^{-x^2} \phi_n(x) \phi_m(x) dx$. This integral must therefore be zero. This means that the functions $\phi_n(x)$ must be orthogonal and therefore multiples of the Hermite polynomials.

If we take $n = m$, the coefficient of $t^n s^n$ on the right-hand side is $2^n \sqrt{\pi}/n!$. Therefore

$$h_n = \int_{-\infty}^{\infty} e^{-x^2} H_n^2(x) dx = 2^n n! \sqrt{\pi}. \tag{2.7.6}$$

2.8 Recurrence Relations

The recurrence relation can be derived using the generating function. If we differentiate Eq. (2.7.5) with respect to t we see that

$$\sum_{n=0}^{\infty} \frac{t^{n-1}}{(n-1)!} H_n(x) = 2(x-t) e^{2xt-t^2} = 2(x-t) \sum_{n=0}^{\infty} \frac{t^n}{n!} H_n(x).$$

Equating the coefficients of t^n on both sides of this equation leads to the recurrence relation:

$$H_{n+1}(x) = 2x H_n(x) - 2n H_{n-1}(x). \tag{2.8.1}$$

If we now differentiate the generating function with respect to x

$$\sum_{n=0}^{\infty} \frac{t^n}{n!} H_n'(x) = \frac{\partial}{\partial x} e^{2xt-t^2} = 2t \sum_{n=0}^{\infty} \frac{t^n}{n!} H_n(x).$$

Equating powers of t on both sides leads to

$$H_n'(x) = 2n H_{n-1}(x). \tag{2.8.2}$$

If we differentiate the generating function m times with respect to x

$$\sum_{n=0}^{\infty} \frac{t^n}{n!} \frac{d^m}{dx^m} H_n(x) = \frac{\partial^m}{\partial x^m} e^{2xt-t^2} = (2t)^m \sum_{n=0}^{\infty} \frac{t^n}{n!} H_n(x).$$

Equating coefficients of powers of t gives

$$\frac{d^m}{dx^m} H_n(x) = \frac{2^m n!}{(n-m)!} H_{n-m}(x),$$

which is the same result as that found in section 2.4.

2.9 Addition Formulae

In this section we find two expressions for $H_n(x + y)$.

$$H_n(x + y) = \sum_{m=0}^{n} \frac{n! \, (2x)^m}{m! \, (n - m)!} H_{n-m}(y)$$

$$= \frac{1}{2^{n/2}} \sum_{m=0}^{n} \frac{n!}{m! \, (n - m)!} H_m(x\sqrt{2}) H_{n-m}(y\sqrt{2}). \tag{2.9.1}$$

Both of these relations can be proved using the generating function.

$$\sum_{n=0}^{\infty} \frac{t^n}{n!} H_n(x + y) = \exp\left(2t(x + y) - t^2\right) = \sum_{p=0}^{\infty} \frac{(2tx)^p}{p!} \sum_{q=0}^{\infty} \frac{t^q}{q!} H_q(y).$$

Equating coefficients of t^n on both sides leads to the first relation of Eq. (2.9.1).

If we now put $t = \sqrt{2}s$ in the generating function,

$$\sum_{n=0}^{\infty} \frac{(\sqrt{2}\,s)^n}{n!} H_n(x + y) = \exp\left(2\sqrt{2}s(x + y) - 2s^2\right)$$

$$= \sum_{p=0}^{\infty} \frac{s^p}{p!} H_p(\sqrt{2}\,x) \sum_{q=0}^{\infty} \frac{s^q}{q!} H_q(\sqrt{2}\,y).$$

Equating the coefficients of s^n on both sides gives the second relation of Eq. (2.9.1).

2.10 Step Up and Step Down Operators

We can use the recurrence relation (2.8.1) to find an expression for $H_{n+1}(x)$ in terms of $H_n(x)$ and its derivative. This is

$$H_{n+1}(x) = S_n^+ H_n(x) = \left\{ 2x - \frac{d}{dx} \right\} H_n(x), \tag{2.10.1}$$

and from (2.8.2),

$$H_{n-1}(x) = S_n^- H_n(x) = \left\{ \frac{1}{2n} \frac{d}{dx} \right\} H_n(x). \tag{2.10.2}$$

2.11   Parabolic Cylinder Functions

The Parabolic Cylinder Function $D_n(x)$ is related to the Hermite polynomial $H_n(x)$ by

$$D_n(x) = \frac{1}{2^{n/2}} e^{-x^2/4} H_n(x/\sqrt{2}).$$ (2.11.1)

$D_n(x)$ satisfies the second order differential equation

$$\frac{d^2u}{dx^2} + \left(n + \frac{1}{2} - \frac{x^2}{4} \right) u = 0.$$ (2.11.2)

Mathematical physicists will recognise this as the equation for the harmonic oscillator in quantum mechanics.

References

Copson E T, Theory of Functions of a Complex Variable, Oxford University Press, 1955.

Courant R and Hilbert D, Methods of Mathematical Physics Vol 1, Interscience Publishers, 1953.

Dennery P and Krzywicki A, Mathematics for Physicists, Harper and Rowe, 1967.

Erdelyi A, Higher Transcendental Functions Vol 2, McGraw-Hill, 1953.

Hochstrasser Urs W, Orthogonal Polynomials, Chapter 22, Handbook of Mathematical Functions, Eds. Abramowitz M and Stegun I A, Dover, 1970.

Koornwinder T H, Wong R, Koekoek R and Swarttouw R F, Chapter 18, NIST Handbook of Mathematical Functions Eds. Olver W J, Lozier D W, Boisvert R F and Clark C W, NIST and Cambridge University Press, 2009.

Miller J C P Parabolic Cylinder Functions, Chapter 19 Handbook of Mathematical Functions, Eds. Abramowitz M and Stegun I A, Dover, 1970.

Pauling L and Wilson E B, Introduction to Quantum Mechanics, McGraw-Hill, 1935.

Schiff L I, Quantum Mechanics, McGraw-Hill, 1955.

Sneddon I N, Special Functions of Mathematical Physics and Chemistry, Oliver and Boyd, 1956.

Temme N M Parabolic Cylinder Functions, Chapter 12 Digital Library of Mathematical Functions NIST 2009.

Titchmarsh E C Introduction to the Theory of Fourier Integrals, Oxford University Press, 1959.

Whittaker E T and Watson G N, Modern Analysis, Cambridge University Press, 1927.

Chapter 3

Associated Laguerre Polynomials

3.1 Introduction

The Associated Laguerre polynomials $L_n^{(\alpha)}(x)$ arise from the orthogonalisation process for polynomials defined over a domain $(0, \infty)$ with a weight function $x^\alpha \exp(-x)$, $\alpha > -1$. When $\alpha = 0$, the polynomials are called Laguerre polynomials $L_n(x)$.

The weight function $x^\alpha e^{-x}$ can be thought of as arising as a limit from the weight function $(x - a)^\alpha (b - x)^\beta$ by firstly putting $a = 0$, dividing through by b and putting $\beta = b$ to obtain $x^\alpha (1 - x/b)^b$ and then taking the limit $b \to \infty$.

If we follow the Gram-Schmidt orthogonalisation procedure described in the first chapter and then multiply the resulting polynomial $R_n(x)$ a number such that the coefficient of x^n is $k_n = (-1)^n/n!$, the result is the Associated Laguerre polynomial $L_n^{(\alpha)}(x)$.

The Associated Laguerre polynomials are multiples of the Confluent Hypergeometric functions:

$$L_n^{(\alpha)}(x) = \frac{\Gamma(\alpha + n + 1)}{\Gamma(\alpha + 1)n!} M(-n, \alpha + 1), x) = \frac{\Gamma(\alpha + n + 1)}{\Gamma(\alpha + 1)n!} {}_1F_1(-n; \alpha + 1; x),$$
(3.1.1)

The first few Laguerre polynomials are

$L_0(x) = 1.$
$L_1(x) = 1 - x.$
$L_2(x) = 1 - 2x + x^2/2.$
$L_3(x) = 1 - 3x + 3x^2/2 - x^3/6.$
$L_4(x) = 1 - 4x + 3x^2 - 2x^3/3 + x^4/24.$
$L_5(x) = 1 - 5x + 5x^2 - 5x^3/3 + 5x^4/24 - x^5/120.$
$L_6(x) = 1 - 6x + 15x^2/2 - 10x^3/3 + 5x^4/8 - x^5/20 + x^6/720.$

Graphs of $e^{-x/2}L_n(x)$ for $n = 1$ to $n = 5$.

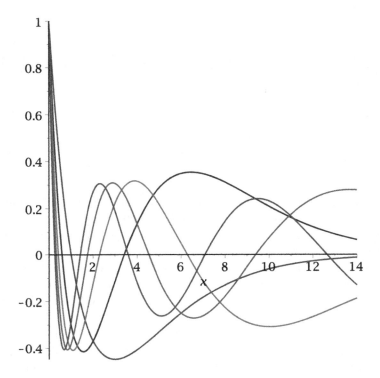

The first 5 Associated Laguerre polynomials are

$L_0^{(\alpha)}(x) = 1.$

$L_1^{(\alpha)}(x) = \alpha + 1 - x.$

$L_2^{(\alpha)}(x) = (\alpha + 1)_2/2 - (\alpha + 2)x + x^2/2.$

$L_3^{(\alpha)}(x) = (\alpha + 1)_3/6 - (\alpha + 2)_2 x/2 + (\alpha + 3)x^2/2 - x^3/6.$

$L_4^{(\alpha)}(x) = (\alpha + 1)_4/24 - (\alpha + 2)_3 x/6 + (\alpha + 3)_2 x^2/4 - (\alpha + 4)x^3/6 + x^4/24.$

In these formulae we have used the Pochammer symbol

$(s)_n = s(s + 1)(s + 2)...(s + n - 1).$

3.2 Differential Equation

In many applications Associated Laguerre polynomials arise as polynomial solutions of a second order ordinary differential equation. We shall examine this differential equation and show that these polynomial solutions are orthogonal with weight factor $x^\alpha \exp(-x)$ and so must be multiples of the Associated Laguerre polynomials.

The differential equation satisfied by Associated Laguerre polynomials is

$$x\frac{d^2y}{dx^2} + (\alpha + 1 - x)\frac{dy}{dx} + \lambda y = 0. \tag{3.2.1}$$

Two linearly independent solutions of this equation are the Confluent Hypergeometric Functions $M(-\lambda, \alpha + 1, x)$ and $U(-\lambda, \alpha + 1, x)$ which is singular at the origin.

The point $x = 0$ is a regular singular point and so we can use the method of Frobenius to find a solution. One solution can be written in the series form $x^c \sum_{n=0}^{\infty} a_n x^n$. If we substitute this into the differential equation, we find

$$\sum_{n=0}^{\infty}(n+c+1)(n+c)a_n x^{n+c-1} + (\alpha+1-x)\sum_{n=0}^{\infty}(n+c)a_n x^{n+c-1} + \lambda\sum_{n=0}^{\infty} a_n x^n = 0.$$

The coefficient of x^{c-1} is $c(c + \alpha)a_0$. By hypothesis, $a_0 \neq 0$ and so $c = 0$ or $c = -\alpha$. The solution with $c = 0$ is a power series. If α has a positive non-integer value we obtain a second power series solution which is singular at the origin. In many practical applications α is a positive integer. In this case the second solution is more complicated but is also singular at the origin. Many applications require a solution which is not singular for any finite value of x. This is the power series solution obtained by taking $c = 0$.

The recurrence relation for the expansion coefficients in the power series for the solution with $c = 0$ is

$$\frac{a_{n+1}}{a_n} = \frac{n - \lambda}{(n + 1)(n + \alpha + 1)}. \tag{3.2.2}$$

If λ is a positive integer, say $\lambda = m$, $a_{m+1}/a_m = 0$, the series terminates and becomes an mth order polynomial. We shall show that these polynomials are orthogonal with weight function $x^{-\alpha}e^{-x}$ and so must be multiples of the Associated Laguerre polynomials $L_n^{(\alpha)}(x)$.

3.3   Orthogonality

Let $R_m(x)$ be the polynomial solution of Eq. (3.2.1) when $\lambda = m$ and $R_n(x)$ be the solution for $\lambda = n$, where m and n are both integers.

If we write the differential equation (3.2.1) for $R_m(x)$ in Sturm Liouville form

$$\frac{d}{dx}\left(x^{\alpha+1} e^{-x} \frac{dR_m(x)}{dx} \right) = -m x^\alpha e^{-x} R_m(x). \tag{3.3.1}$$

Multiply this by $R_n(x)$, where $n \neq m$ and integrate from 0 to ∞. The left-hand side becomes

$$\int_0^\infty R_n(x) \frac{d}{dx}\left(x^{\alpha+1} e^{-x} \frac{dR_m(x)}{dx} \right) dx.$$

On integrating by parts and noting that for $\alpha > -1$ the integrated part vanishes at each integration limit, we obtain

$$\int_0^\infty x^{\alpha+1} e^{-x} \frac{dR_n(x)}{dx} \frac{dR_m(x)}{dx} dx = m \int_0^\infty x^\alpha e^{-x} R_n(x) R_m(x) dx.$$

If we now follow the procedure above with n and m interchanged we arrive at an equation which is the same as that above except that the integral on the right is multiplied by n instead of m. If we take the difference of these two equations, we get

$$(m-n) \int_0^\infty x^\alpha e^{-x} R_n(x) R_m(x) dx = 0.$$

Since $m \neq n$ it follows that

$$\int_0^\infty x^\alpha e^{-x} R_n(x) R_m(x) dx = 0. \tag{3.3.2}$$

This is the orthogonality relation satisfied by the Associated Laguerre polynomials. The functions $R_n(x)$ must therefore be multiples of the Associated Laguerre polynomials. If we multiply $R_n(x)$ by a constant such that the coefficient of x^n is $(-1)^n/n!$, we will obtain the Associated Laguerre polynomial $L_n^{(\alpha)}(x)$.

If we use the recurrence relation (3.2.2) for the expansion coefficients, we find that the coefficient of x^{n-1},

$$k_n' = (-1)^{n-1}(n+\alpha)/(n-1)!. \tag{3.3.3}$$

3.4 Derivative Property

The pth derivative of $L_m^{(\alpha)}(x)$ is a polynomial of order $m - p$. We now show that it is a multiple of $L_{m-p}^{(\alpha+p)}(x)$. If we differentiate the differential equation for Associated Laguerre polynomials

$$x\frac{d^2y}{dx^2} + (1 + \alpha - x)\frac{dy}{dx} + my = 0 \tag{3.4.1}$$

p times, we obtain

$$x\frac{d^2y}{dx^2} + (1 + \alpha + p - x)\frac{dy}{dx} + (m - p)y = 0. \tag{3.4.2}$$

It follows that a solution of Eq. (3.4.2) is the pth derivative of a solution of Eq. (3.4.1). The polynomial solution of Eq. (3.4.2), $L_{m-p}^{(\alpha+p)}(x)$ is therefore a multiple of the pth derivative of $L_m^{(\alpha)}(x)$. We can find the constant by considering the highest powers of x. The coefficient of the highest power of x in the pth derivative of $L_m^{(\alpha)}(x)$ is

$$\frac{(-1)^m}{m!}\frac{m!}{(m-p)!} = \frac{(-1)^m}{(m-p)!}.$$

The coefficient of x^{m-p} in $L_{m-p}^{(\alpha+p)}(x)$ is $(-1)^{m-p}/(m-p)!$. The constant is therefore $(-1)^p$.

$$\frac{d^p}{dx^p}L_m^{(\alpha)}(x) = (-1)^p L_{m-p}^{(\alpha+p)}(x). \tag{3.4.3}$$

3.5 Rodrigues Formula

We firstly note that if $m > n$ and $\alpha > -1$ then

$$\frac{d^n}{dx^n}\left(x^{\alpha+m}e^{-x}\right) = 0 \qquad \text{when} \qquad x = 0,$$

and tends to zero as $x \to \infty$. From this result we deduce that

$$\int_0^\infty \frac{d^n}{dx^n}\left(e^{-x}x^{\alpha+n}\right)dx = 0,$$

and using the method of integration by parts we see that

$$\int_0^\infty x\frac{d^n}{dx^n}\left(e^{-x}x^{\alpha+n}\right)dx = 0 \qquad n > 1.$$

If we integrate by parts m times we see that

$$\int_0^\infty x^m\frac{d^n}{dx^n}\left(e^{-x}x^{\alpha+n}\right)dx = 0 \qquad \text{provided that} \qquad n > m. \tag{3.5.1}$$

In other words, the nth order polynomial

$$Q_n(x) = x^{-\alpha} e^x \frac{d^n}{dx^n} \left(e^{-x} x^{n+\alpha} \right)$$

is orthogonal to x^m for all values of $n > m$. This means that $Q_m(x)$ is orthogonal to $Q_n(x)$ for all values of $n > m$. These polynomials must therefore be multiples of the Associated Laguerre polynomials we found earlier.

We see that the coefficient of x^n in the polynomial $Q_n(x)$ is $(-1)^n$. Therefore

$$L_n^{(\alpha)}(x) = \frac{1}{n!} x^{-\alpha} e^x \frac{d^n}{dx^n} \left(e^{-x} x^{\alpha+n} \right). \tag{3.5.2}$$

This can be put into a symbolic form by moving e^{-x} through the derivatives

$$L_n^{\alpha}(x) = \frac{1}{n!} x^{-\alpha} \left(\frac{d}{dx} - 1 \right)^n x^{n+\alpha}. \tag{3.5.3}$$

Another differential representation for the Associated Laguerre polynomial is Eq. (H15) derived in the General Appendix:

$$L_n^{\alpha}(x) = \frac{(-1)^n}{n!} \frac{d^n}{dt^n} \left\{ (1-t)^{\alpha+n} e^{xt} \right\} \bigg|_{t \to 0}. \tag{3.5.4}$$

3.6   Explicit Expression

Evaluating the derivative in (3.5.2) leads to

$$L_n^{(\alpha)}(x) = \sum_{m=0}^{n} \frac{\Gamma(n+\alpha+1)}{\Gamma(m+\alpha+1)(n-m)!\, m!} (-x)^m. \tag{3.6.1}$$

We have seen that the coefficient of x^n in $L_n^{(\alpha)}(x)$ is $k_n = (-1)^n/n!$ and the coefficient of x^{n-1} is $k_n' = (-1)^{n-1}(n+\alpha)/(n-1)!$.

We can use (3.6.1) to define $L_n^{(m)}(x)$ for negative integer values of m. We see that

$$L_n^{(-p)}(x) = \frac{(n-p)!}{n!} (-x)^p L_{n-p}^{(p)}(x). \tag{3.6.2}$$

We can express x^n as a linear combination of Laguerre polynomials

$$x^n = \sum_{m=0}^{n} b_m L_m^{(\alpha)}(x). \tag{3.6.3}$$

The coefficients b_m can be evaluated by multiplying Eq. (3.6.3) by $x^\alpha e^{-x} L_k^{(\alpha)}(x)$ and integrating from 0 to ∞. Since the Laguerre polynomials form an orthogonal set, we have

$$\int_0^\infty x^\alpha e^{-x} L_m^{(\alpha)}(x) x^n dx = b_m \int_0^\infty x^\alpha \left[L_m^{(\alpha)}(x) \right]^2 dx = b_m h_m. \qquad (3.6.4)$$

The integral on the left can be evaluated using Rodrigues formula. It becomes

$$\frac{1}{m!} \int_0^\infty x^n \frac{d^m}{dx^m} \left(e^{-x} x^{\alpha+m} \right) dx = \frac{(-1)^m n!}{m!(n-m)!} \int_0^\infty e^{-x} x^{\alpha+n} dx$$

on integrating by parts m times. Using the value of h_m from Eq. (3.7.6) below we find

$$x^n = \sum_{m=0}^n \frac{(-1)^m n! \, \Gamma(\alpha+n+1)}{(n-m)! \, \Gamma(\alpha+m+1)} L_m^{(\alpha)}(x). \qquad (3.6.5)$$

3.7   Generating Function

If we define $w(x,t)$ by

$$w(x,t) = \frac{1}{(1-t)^{\alpha+1}} \exp\left(\frac{xt}{t-1} \right),$$

we can show that

$$x \frac{\partial^2 w}{\partial x^2} + (\alpha + 1 - x) \frac{\partial w}{\partial x} = -t \frac{\partial w}{\partial t}. \qquad (3.7.1)$$

This is most easily done using a computer algebra package such as Maple. Let us define $\phi_n(x)$ by

$$\sum_{n=0}^\infty t^n \phi_n(x) = \frac{1}{(1-t)^{\alpha+1}} \exp\left(\frac{xt}{t-1} \right) \qquad |t| < 1. \qquad (3.7.2)$$

The functions $\phi_n(x)$ are nth order polynomials in x. We can see this by expanding the exponential and writing w as

$$w(x,t) = \sum_{n=0}^\infty (-x)^n t^n \text{ times the polynomial expansion of } (1-t)^{\alpha+n+1}/n!.$$

On substituting the left-hand side of (3.7.2) into (3.7.1), it follows that

$$\sum_{n=0}^\infty t^n \left\{ x \frac{d^2 \phi_n}{dx^2} + (\alpha + 1 - x) \frac{d\phi_n}{dx} \right\} = -t \sum_{n=0}^\infty \phi_n(x) \frac{d}{dt} t^n = - \sum_{n=0}^\infty n t^n \phi_n(x). \qquad (3.7.3)$$

If we equate the coefficients of powers of t on both sides of the equation we see that

$$x\frac{d^2}{dx^2}\phi_n(x) + (\alpha + 1 - x)\frac{d}{dx}\phi_n(x) = -n\phi_n(x). \qquad (3.7.4)$$

In other words $\phi_n(x)$ satisfies the Associated Laguerre equation and is therefore some multiple of the Associated Laguerre polynomial $L_n^{(\alpha)}(x)$.

If we examine the expansion for $\phi_n(x)$, we see that the coefficient of x^n is $(-1)^n/n!$ and so $\phi_n(x) = L_n^{(\alpha)}(x)$.

The generating function for the Associated Laguerre polynomials is then

$$w(x,t) = \sum_{n=0}^{\infty} t^n L_n^{(\alpha)}(x) = \frac{1}{(1-t)^{\alpha+1}} \exp\left(\frac{xt}{t-1}\right) \qquad |t| < 1. \qquad (3.7.5)$$

Let us now evaluate the integral from 0 to ∞ of $w(x,t)w(x,s)$ multiplied by the weight function $x^\alpha \exp(-x)$. Expanding the functions $w(x,t)$ and $w(x,s)$ in powers of t and s gives

$$\sum_{n,m=0}^{\infty} t^n s^m \int_0^\infty x^\alpha e^{-x} L_n^{(\alpha)}(x) L_m^{(\alpha)}(x) dx$$

$$= \int_0^\infty x^\alpha e^{-x} \frac{1}{(1-t)^{\alpha+1}} \exp\left(\frac{xt}{t-1}\right) \frac{1}{(1-s)^{\alpha+1}} \exp\left(\frac{xs}{s-1}\right) dx$$

$$= \int_0^\infty x^\alpha \frac{1}{(1-t)^{\alpha+1}} \frac{1}{(1-s)^{\alpha+1}} \exp\left(\frac{-x(1-st)}{(1-t)(1-s)}\right) dx$$

$$= \frac{1}{(1-st)^{\alpha+1}} \int_0^\infty v^\alpha e^{-v} dv = \frac{\Gamma(\alpha+1)}{(1-st)^{\alpha+1}}.$$

This expression has an expansion of the form $\sum_{n=0}^{\infty} a_n(ts)^n$. This means that the coefficient of $t^n s^m$ for $n \neq m$ is zero. This confirms the orthogonality relation for the polynomials.

We can use this result to evaluate the integral when $n = m$.

$$h_n = \int_0^\infty x^\alpha e^{-x} \left[L_n^{(\alpha)}(x)\right]^2 dx$$

$$= \text{ the coefficient of } (st)^n \text{ in the expansion of } \frac{\Gamma(\alpha+1)}{(1-st)^{\alpha+1}},$$

that is

$$h_n = \int_0^\infty x^\alpha e^{-x} (L_n^{(\alpha)}(x))^2 dx = \Gamma(\alpha + n + 1)/n!. \qquad (3.7.6)$$

3.8 Recurrence Relations

A number of recurrence relations can be derived from the generating function. We can show that

$$(1-t)^2 \frac{\partial}{\partial t} \left\{ \frac{1}{(1-t)^{\alpha+1}} \exp\left(\frac{xt}{t-1}\right) \right\}$$

$$= \left[(\alpha+1)(1-t) - x \right] \frac{1}{(1-t)^{\alpha+1}} \exp\left(\frac{xt}{t-1}\right).$$

This is most easily done using a computer algebra package. Substituting the generating function expansion leads to:

$$(1-t)^2 \sum_{n=1}^{\infty} nt^{n-1} L_n^{(\alpha)}(x). = \left[(\alpha+1)(1-t) - x \right] \sum_{n=0}^{\infty} t^n L_n^{(\alpha)}(x).$$

Equating coefficients of t^n on both sides of this equation leads to the recurrence relation

$$(n+1)L_{n+1}^{(\alpha)}(x) = (2n+\alpha+1-x)L_n^{(\alpha)}(x) - (n+\alpha)L_{n-1}^{(\alpha)}(x). \quad (3.8.1)$$

Again

$$t\frac{\partial}{\partial t} \left\{ \frac{1}{(1-t)^{\alpha}} \exp\left(\frac{xt}{t-1}\right) \right\} = \left\{ \frac{\alpha t}{(1-t)^{\alpha+1}} - \frac{xt}{(1-t)^{\alpha+2}} \right\} \exp\left(\frac{xt}{t-1}\right).$$

This means that

$$\sum_{n=0}^{\infty} \left[nt^n - (n+1)t^{n+1} \right] L_n^{(\alpha)}(x) = \alpha t \sum_{n=0}^{\infty} t^n L_n^{(\alpha)}(x) - xt \sum_{n=0}^{\infty} t^n L_n^{\alpha+1}(x).$$

Equating coefficients of powers of t^n leads to

$$n\left[L_n^{(\alpha)}(x) - L_{n-1}^{(\alpha)}(x) \right] = \alpha L_{n-1}^{(\alpha)}(x) - x L_{n-1}^{(\alpha+1)}(x). \quad (3.8.2)$$

Combining Eq. (3.8.1) and Eq. (3.8.2) gives

$$x L_n^{(\alpha)+1}(x) = (n+\alpha)L_{n-1}^{(\alpha)}(x) - (n-x)L_n^{(\alpha)}(x). \quad (3.8.3)$$

Another relation which can easily be proved using the generating function is

$$L_n^{(\alpha)}(x) = L_n^{\alpha+1}(x) - L_{n-1}^{\alpha+1}(x) = \sum_{j=0}^{k} \frac{k!}{(k-j)!\,j!} L_{n-j}^{(\alpha-k+j)}(x). \quad (3.8.4)$$

The first part is proved by equating the coefficients of powers of t on both sides of

$$\sum_{n=0}^{\infty} t^n L_n^{(\alpha)}(x) = \frac{1}{(1-t)^{\alpha+1}} \exp\left(\frac{xt}{t-1}\right)$$

$$= \frac{1-t}{(1-t)^{\alpha+2}} \exp\left(\frac{xt}{t-1}\right) = (1-t) \sum_{n=0}^{\infty} t^n L_n^{(\alpha+1)}(x).$$

The other part can be proved by noting firstly that

$$\sum_{j=0}^{k} \frac{k!}{(k-j)! \, j!} \frac{t^j}{1-t)^{j-k}} = 1.$$

Hence

$$\sum_{n=0}^{\infty} t^n L_n^{(\alpha)}(x) = \sum_{j=0}^{k} \frac{k!}{(k-j)! \, j!} \frac{t^j}{(1-t)^{\alpha+j-k+1}} \exp\left(\frac{xt}{t-1}\right)$$

$$= \sum_{j=0}^{k} \frac{k!}{(k-j)! \, j!} \sum_{n=0}^{\infty} t^{n+j} L_n^{\alpha+j-k}(x).$$

Equating the coefficients of powers of t on both sides gives the second part of Eq. (3.8.4).

Substituting $L_n^{(\alpha)}(x) = L_n^{(\alpha+1)}(x) - L_{n-1}^{(\alpha+1)}(x)$ into Eq. (3.8.2) gives

$$nL_n^{(\alpha+1)}(x) = (n-x)L_{n-1}^{(\alpha+1)}(x) + (n+\alpha)L_{n-1}^{(\alpha)}(x). \tag{3.8.5}$$

There are some more relations that can be proved by direct substitution:

$$L_n^{(\alpha)}(x) = \sum_{m=0}^{n} \frac{\Gamma(\alpha-\beta+n+1)}{\Gamma(\alpha-\beta+m+1)(n-m)!} L_m^{(\beta-m)}(x). \tag{3.8.6}$$

Substituting the series expansion of $L_m^{(\beta-m)}(x)$ gives

$$\sum_{m=0}^{n} \frac{\Gamma(\alpha-\beta+n+1)}{\Gamma(\alpha-\beta+m+1)(n-m)!} \sum_{k=0}^{m} \frac{\Gamma(\beta+1)}{\Gamma(k+\beta-m+1)(m-k)! \, k!} (-x)^k.$$

Interchanging the order of the summations and replacing $m-k$ by m leads to

$$\sum_{k=0}^{n} \frac{(-x)^k}{k!} \sum_{m=0}^{n-k} \frac{\Gamma(\alpha-\beta+n+1)\Gamma(\beta+1)}{\Gamma(\alpha-\beta+k+m+1)(n-k-m)!\Gamma(\beta-m+1)m!}$$

$$= \sum_{k=0}^{n} \frac{(-x)^k}{k!} \frac{\Gamma(\alpha-\beta+n+1)}{\Gamma(\alpha-\beta+k+1)(n-k)!} \left\{ 1 + \frac{\beta(n-k)}{(\alpha-\beta+k+1).1} + \cdots \right\}$$

$$= \sum_{k=0}^{n} \frac{(-x)^k}{k!} \frac{\Gamma(\alpha-\beta+n+1)}{\Gamma(\alpha-\beta+k+1)(n-k)!} \, _2F_1(-\beta, k-n; \alpha-\beta+k+1; 1)$$

$$= \sum_{k=0}^{n} \frac{\Gamma(\alpha + n + 1)}{\Gamma(\alpha + k + 1)(n - k)! k!} (-x)^k = L_n^{(\alpha)}(x),$$

where we have used Vandermonde's theorem for the hypergeometric function $_2F_1(a, -n; c; 1) = (c - a)_n / c_n$, (see the General Appendix).

We also have:

$$\frac{x^k}{k!} L_n^{(\alpha)}(x) = \sum_{i=0}^{k} (-1)^i \frac{(n + i)! \, \Gamma(n + \alpha + 1)}{n! \, i! \, \Gamma(n + \alpha + i - k + 1)(k - i)!} L_{n+i}^{(\alpha-k)}(x). \quad (3.8.7)$$

Substituting the expansion for the Laguerre polynomial on the right-hand side

$$\sum_{i=0}^{k} (-1)^i \frac{(n + i)! \, \Gamma(n + \alpha + 1)}{n! \, i! \, (k - i)!} \sum_{j=0}^{n+i} \frac{(-x)^j}{\Gamma(\alpha - k + j + 1)(n + i - j)! \, j!}.$$

If $j \leq n$, the coefficient of $(-x)^j$ is

$$\sum_{i=0}^{k} \frac{(-1)^i (n + i)! \, \Gamma(\alpha + n + 1)}{n! \, i! \, (k - i)! \, \Gamma(\alpha - k + j + 1)(n + i - j)! \, j!}$$

$$= \frac{\Gamma(\alpha + n + 1)}{k! \, \Gamma(\alpha - k + j + 1)(n - j)! \, j!} \left\{ 1 - \frac{k(n + 1)}{n - j + 1} + \frac{k(k - 1)(n + 1)(n + 2)}{(n - j + 1)(n - j + 2)2!} \cdots \right\}.$$

We use Vandermonde's Theorem (see the General Appendix) to evaluate the sum in the curly brackets $_2F_1(n + 1, -k; n - j + 1; 1)$.

$$_2F_1(n + 1, -k; n - j + 1; 1) = \frac{(-j)_k}{(n + 1 - j)_k}.$$

We note that $(-j)_k = (-j)(-j + 1)...(j + k - 1)$. If $j < k$, one of these factors will be zero. There will therefore be no terms with powers of x less than k. For $j \geq k$, $(-j)_k = (-1)^k j! / (j - k)!$. Therefore the coefficient of x^j is

$$= \frac{\Gamma(\alpha + n + 1)(-1)^{j-k}}{k! \, \Gamma(\alpha + j - k + 1)(n + k - j)! \, (j - k)!},$$

which is the coefficient of x^j in $x^k L_n^{(\alpha)}(x)/k!$.

If $j \geq n$, the coefficient of $(-x)^j$ is

$$\sum_{i=j-n}^{k} \frac{(-1)^i (n + i)! \, \Gamma(\alpha + n + 1)}{n! \, i! \, (k - i)! \, \Gamma(\alpha - k + j + 1)(n + i - j)! \, j!}$$

$$= \frac{(-1)^{j-n} \Gamma(n + \alpha + 1)}{\Gamma(\alpha - k + j + 1) n! \, (j - n)! (k - j + n)!} \left\{ 1 - \frac{(j + 1)(k - j + n)}{(j - n + 1)} + ... \right\}.$$

Using Vandermonde's Theorem on the sum in the curly brackets

$$_2F_1(j+1,j-k-n;j-n+1;1) = \frac{(-n)_{n+k-j}}{(j-n+1)_{n+k-j}}.$$

We note again that if $j < k$ there will be a factor of zero in $(-n)_{n+k-j}$ and so again no powers of x less than x^k. For $j \geq k$ we get the same coefficient of x^j as before. This establishes Eq. (3.8.7).

Another pair of relations is

$$L_n^{(\alpha)}(x) - \sum_{j=0}^{m-1} \frac{\Gamma(n+\alpha+1)(-x)^j}{\Gamma(\alpha+j+1)(n-j)!\,j!}$$

$$= \frac{(-x)^m}{(m-1)!\,n!} \sum_{i=0}^{n-m} \frac{\Gamma(\alpha+n+1)\,i!\,(n-i-1)!}{\Gamma(\alpha+m+i+1)(n-m-i)!} L_i^{(\alpha+m)}(x) \qquad (3.8.8)$$

$$= \frac{(-x)^m}{(m-1)!\,n!} \sum_{i=0}^{n-m} \frac{\Gamma(n+\alpha-i)(n-i-1)!\,i!}{\Gamma(\alpha+m)(n-m-i)!} L_i^{(n+\alpha+m-i)}(x). \qquad (3.8.9)$$

The part on the left of Eq. (3.8.8) is just the terms in the series expansion of $L_n^{(\alpha)}(x)$ with powers of x^k with $k \geq m$. Eq. (3.8.8) will be shown to be true if

$$\sum_{k=m}^{n} \frac{(-x)^k}{\Gamma(\alpha+k+1)(n-k)!\,k!}$$

$$= \frac{(-x)^m}{(m-1)!\,n!} \sum_{i=0}^{n-m} \frac{i!\,(n-i-1)!}{(n-m-i)!} \sum_{k=0}^{i} \frac{(-x)^k}{\Gamma(\alpha+m+k+1)(i-k)!\,k!}.$$

$$(3.8.10)$$

Interchanging the order of the summations on the right-hand side and replacing $i - k$ by i, we find the coefficient of $(-x)^{k+m}$ is

$$\sum_{i=0}^{n-k-m} \frac{(i+k)!\,(n-i-k-1)!}{(m-1)!\,n!\,\Gamma(\alpha+m+k+1)\,(n-m-i-k)!\,i!\,k!}$$

$$= \frac{(n-k-1)!}{(m-1)!\,n!\,\Gamma(\alpha+m+k+1)(n-m-k)!} \left\{ 1 + \frac{(k+1)(n-m-k)}{(n-k-1)} + \ldots \right\}.$$

The sum inside the curly brackets is the hypergeometric polynomial $_2F_1(k+1, m+k-n; k+1-n)$ which is most easily evaluated using Vandermonde's theorem

$$_2F_1(k+1, m+k-n; k+1-n; 1) = \frac{(-n)_{n-m-k}}{k+1-n)_{n-m-k}}$$

$$= \frac{(k+1+m)_{n-m-k}}{m_{n-m-k}} = \frac{n!\,(m-1)!}{(k+m)!(n-k-1)!}.$$

The coefficient of $(-x)^{m+k}$ on the right-hand side of Eq. (3.8.10) becomes $1/[\Gamma(\alpha+m+k+1)(n-k-m)!(k+m)!]$ which is the same as that on the left-hand side.

We now treat the right-hand side of Eq. (3.8.9) in the same way as for Eq. (3.8.8). Expanding the Laguerre polynomial and interchanging the orders of summation produces

$$\sum_{k=0}^{n-m} A_k \frac{(-x)^{k+m}}{k!\,(m-1)!}$$

where

$$A_k = \sum_{i=k}^{n-m} \frac{\Gamma(n+\alpha-i)(n-i-1)!\,i!\,\Gamma(n+\alpha+m+1)}{n!\,\Gamma(\alpha+m)(n-m-i)!\,\Gamma(n+\alpha+m-i+k+1)(i-k)!}.$$

The coefficient of $(-x)^{k+m}$ on the right-hand side of Eq. (3.8.9) is then

$$\frac{\Gamma(n+\alpha-k)(n-k-1)!}{n!\,(m-1)!\,\Gamma(\alpha+m)(n-m-k)!}S.$$

Where

$$S = 1 + \frac{(k+1)(n-m-k)(n+\alpha+m)}{(n+\alpha-k-1)(n-k-1)} + \cdots \quad .$$

This sum is the generalised hypergeometric polynomial

$$_3F_2(k+1, -\alpha-n-m, -n; k+1-n-\alpha, k+1-n; 1)$$

$$= \frac{(-n)_{n-k-m}(\alpha+m+k+1)_{n-k-m}}{(k+1-n)_{n-k-m}(\alpha+m)_{n-k-m}}$$

$$= \frac{n!\,(m-1)!\,\Gamma(\alpha+n+1)\Gamma(\alpha+m)}{(m+k)!\,(n-k-1)!\,\Gamma(\alpha+m+k+1)\Gamma(\alpha+n-k)}$$

which we have evaluated using Saalchutz's theorem, see the General Appendix. Combining this with the term in front of the curly brackets gives the coefficient of $(-x)^{k+m}$, $\Gamma(n+\alpha+1)/\Gamma(\alpha+k+1)(n-k-m)!(k+m)!$. This completes the proof of Eq. (3.8.9).

3.9   Addition Formulae

The generating function can be used to prove the identity

$$L_n^{(\alpha+\beta+1)}(x+y) = \sum_{m=0}^{n} L_m^{(\alpha)}(x)L_{n-m}^{(\beta)}(y). \qquad (3.9.1)$$

The generating function for the Laguerre polynomial on the left-hand side

$$\frac{1}{(1-t)^{\alpha+\beta+2}} \exp\left(\frac{t(x+y)}{t-1}\right)$$

$$= \frac{1}{(1-t)^{\alpha+1}} \exp\left(\frac{tx}{t-1}\right) \frac{1}{(1-t)^{\beta+1}} \exp\left(\frac{ty}{t-1}\right).$$

Substituting the expansion Eq. (3.7.5) gives

$$\sum_{n=0}^{\infty} t^n L_n^{(\alpha+\beta+1)}(x+y) = \sum_{m=0}^{\infty} t^m L_n^{(\alpha)}(x) \sum_{p=0}^{\infty} t^p L_n^{(\beta)}(y).$$

Equating coefficients of t^n on both sides of this equation leads directly to Eq. (3.9.1).

Equation (3.9.1) can be generalised to a sum of m arguments

$$L_n^{(\alpha_1+\alpha_2+..\alpha_m+m-1)}(x) = \sum_{p_1+p_2+..p_m=n} L_{p_1}^{(\alpha_1)}(x)L_{p_2}^{(\alpha_2)}(x)...L_{p_m}^{(\alpha_m)}(x).$$

$$(3.9.2)$$

The proof of this relation is an extension of the method used for Eq. (3.9.1).

If we now put $\beta = 0$ and $y = 0$ in Eq. (3.9.1), we obtain

$$L_n^{(\alpha+1)}(x) = \sum_{m=0}^{n} L_m^{(\alpha)}(x). \qquad (3.9.3)$$

If in Eq. (3.9.1) we replace α by $\alpha - \beta - 1$, put $x = 0$ and then replace y by x, we get

$$L_n^{(\alpha)}(x) = \sum_{m=0}^{n} L_m^{(\alpha-\beta-1)}(0)L_{n-m}^{(\beta)}(x) = \sum_{m=0}^{n} \frac{\Gamma(\alpha-\beta+m)}{\Gamma(\alpha-\beta)m!}L_{n-m}^{(\beta)}(x).$$

$$(3.9.4)$$

Another relation that can be derived from the generating function Eq. (3.7.5) is

$$L_n^{(\alpha)}(x) = \sum_{m=0}^{n} L_{n-m}^{(\alpha+m)}(y)\frac{(y-x)^m}{m!}. \qquad (3.9.5)$$

Consider

$$\sum_{m=0}^{\infty} \frac{(y-x)^m}{m!} \sum_{n=0}^{\infty} t^{n+m} L_n^{(\alpha+m)}(y) = \sum_{m=0}^{\infty} \frac{t^m(y-x)^m}{m!\,(1-t)^{\alpha+m+1}} \exp\left(\frac{ty}{t-1}\right)$$

$$= \frac{1}{(1-t)^{\alpha+1}} \exp\left(\frac{ty}{t-1}\right) \exp\left(\frac{t(x-y)}{t-1}\right) = \frac{1}{(1-t)^{\alpha+1}} \exp\left(\frac{tx}{t-1}\right).$$

Substituting the Laguerre polynomial expansion and equating powers of t gives Eq. (3.9.5).

3.10 Differential Relations

We can show that

$$x \frac{\partial}{\partial x} \left\{ \frac{1}{(1-t)^{\alpha+1}} \exp\left(\frac{xt}{t-1}\right) \right\}$$

$$= t \frac{\partial}{\partial t} \left\{ \frac{1}{(1-t)^{\alpha+1}} \exp\left(\frac{xt}{t-1}\right) \right\} - t^{1-\alpha} \frac{\partial}{\partial t} \left\{ \frac{t^{\alpha+1}}{(1-t)^{\alpha+1}} \exp\left(\frac{xt}{t-1}\right) \right\}.$$

Substituting the polynomial expansions into the generating functions gives

$$\sum_{n=0}^{\infty} x t^n \frac{dL_n^{(\alpha)}(x)}{dx} = \sum_{n=1}^{\infty} n t^n L_n^{(\alpha)}(x) - \sum_{n=0}^{\infty} (n+\alpha+1) t^{n+1} L_n^{(\alpha)}(x).$$

If we equate the coefficients of t^n on both sides of this equation we see that

$$x \frac{dL_n^{(\alpha)}(x)}{dx} = n L_n^{(\alpha)}(x) - (n+\alpha) L_{n-1}^{(\alpha)}(x). \tag{3.10.1}$$

The generating function can also be used to derive Eq. (3.4.3). Consider

$$\frac{\partial^p}{\partial x^p} \left\{ \frac{1}{(1-t)^{\alpha+1}} \exp\left(\frac{xt}{t-1}\right) \right\} = \frac{(-1)^p \, t^p}{(1-t)^{\alpha+p+1}} \exp\left(\frac{xt}{t-1}\right).$$

Substituting the expansion of the Associated Laguerre polynomials

$$\sum_{n=0}^{\infty} t^n \frac{d^p}{dx^p} L_n^\alpha(x) = (-1)^p \sum_{n=0}^{\infty} t^{n+p} L_n^{\alpha+p}(x).$$

Equating coefficients of t^n on both sides of this equation gives us (3.4.3).

3.11 Step Up and Step Down Operators

We can use the recurrence relation (3.10.1) to find a formula for $L_{n-1}^{(\alpha)}(x)$ in terms of $L_n^{(\alpha)}(x)$ and its derivative. We see that

$$L_{n-1}^{(\alpha)}(x) = S_n^- L_n^{(\alpha)}(x) = \frac{1}{n+\alpha} \left\{ n - x \frac{d}{dx} \right\} L_n^{(\alpha)}(x). \tag{3.11.1}$$

Combining this with relation (3.8.1), we get

$$L_{n+1}^{(\alpha)}(x) = S_n^+ L_n^{(\alpha)}(x) = \frac{1}{n+1} \left\{ (n+\alpha+1-x) + x \frac{d}{dx} \right\} L_n^{(\alpha)}(x). \tag{3.11.2}$$

References

Copson E T, Theory of Functions of a Complex Variable, Oxford University Press 1955.

Courant R and Hilbert D, Methods of Mathematical Physics Vol 1, Interscience Publishers, 1953.

Dennery P and Krzywicki A, Mathematics for Physicists, Harper and Rowe, 1967.

Hochstrasser Urs W, Orthogonal Polynomials, Chapter 22, Handbook of Mathematical Functions, Eds. Abramowitz M and Stegun I A, Dover, 1970.

Koornwinder T H, Wong R, Koekoek R and Swarttouw R F, Chapter 18, NIST Handbook of Mathematical Functions, Eds. Olver W J, Lozier D W, Boisvert R F and Clark C W, NIST and Cambridge University Press, 2009.

Pauling L and Wilson E B, Introduction to Quantum Mechanics, McGraw-Hill, 1935.

Schiff L I, Quantum Mechanics, McGraw-Hill, 1955.

Sneddon I N, Special Functions of Mathematical Physics and Chemistry, Oliver and Boyd 1956.

Chapter 4

Legendre Polynomials

4.1 Introduction

Legendre polynomials $P_n(x)$ arise from the orthogonalisation process for polynomials in the domain $(-1, 1)$ with a weight factor 1. They were the first set of orthogonal polynomials to be described.

They are a special case of the Jacobi polynomials with α and β both equal to zero. They are also equal to the Gegenbauer polynomials $C_n^{(1/2)}(x)$. That is

$$P_n(x) = C_n^{(1/2)}(x) = P_n^{(0,0)}(x). \qquad (4.1.1)$$

Legendre polynomials can be obtained using the Gram-Schmidt orthogonalisation process described in the first chapter with a weight factor $w(x) = 1$ and then multiplying each of the resulting polynomials by a number such that its value at $x = 1$ is 1.

The first few Legendre polynomials are
$P_0(x) = 1.$
$P_1(x) = x.$
$P_2(x) = (3x^2 - 1)/2.$
$P_3(x) = (5x^3 - 3x)/2.$
$P_4(x) = (35x^4 - 30x^2 + 3)/8.$
$P_5(x) = (63x^5 - 70x^3 + 15x)/8.$
$P_6(x) = (231x^6 - 315x^4 + 105x^2 - 5)/16.$
$P_7(x) = (429x^7 - 693x^5 + 315x^3 - 35x)/16.$
$P_8(x) = (6435x^8 - 12012x^6 + 6930x^4 - 1260x^2 + 35)/128.$
$P_9(x) = (12155x^9 - 25740x^7 + 18018x^5 - 4620x^3 + 315x)/128.$
$P_{10}(x) = (46189x^{10} - 109395x^8 + 90090x^6 - 30030x^4 + 3465x^2 - 63)/256.$

Graphs of $P_n(x)$ for $n = 1$ to $n = 6$.

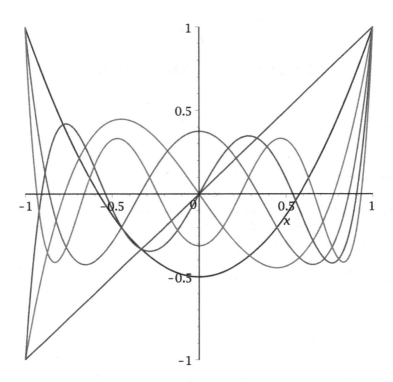

4.2   Differential Equation

Legendre polynomials also arise from the non-singular solution of the ordinary differential equation:

$$(1 - x^2)\frac{d^2y}{dx^2} - 2x\frac{dy}{dx} + \lambda y = 0. \tag{4.2.1}$$

The point $x = 0$ is an ordinary point. This means that we can express the solution in the form of a power series, $y = \sum_{n=0}^{\infty} a_n x^n$. On substituting this power series into the differential equation, we find

$$(1 - x^2)\sum_{n=0}^{\infty} n(n-1)a_n x^{n-2} - 2x\sum_{n=0}^{\infty} na_n x^{n-1} + \lambda \sum_{n=0}^{\infty} a_n x^n.$$

If we equate the coefficients of x^n to zero we get

$$(n+2)(n+1)a_{n+2} + [\lambda - n(n+1)]a_n = 0$$

leading to the recurrence relation

$$\frac{a_{n+2}}{a_n} = \frac{n(n+1) - \lambda}{(n+2)(n+1)}. \tag{4.2.2}$$

We see that there are two series solutions, one series contains only even powers of x, the other is an odd power series.

We can use the ratio test to show that these series solutions both converge for $|x| < 1$, but since $\lim_{n \to \infty}(a_{n+2}/a_n) = 1$, the test fails for $x = 1$. In many applications we require a solution which is finite for $|x| = 1$. To test for convergence at $x = 1$ we can use Gauss's test (see e.g. Phillips, p. 129). Writing $u_m = a_{2m}$,

$$\frac{u_m}{u_{m+1}} = \frac{2m+2)(2m+1)}{2m(2m+1) - \lambda} = 1 + \frac{1}{m} + O\left(\frac{1}{m^2}\right).$$

The coefficient of $1/m$ in the ratio for successive terms is thus equal to 1 and therefore the series diverges. The same argument applies to the series of odd terms. This means that the solution will only be finite at $|x| = 1$ if the series terminates.

We see from Eq. (4.2.2) that one of the series solutions will terminate if $\lambda = m(m+1)$ for some positive integer m. The case when m is a negative integer leads to the same recurrence relation and hence the same solution. If m is a positive even number, the series of even powers will terminate but the odd series will not, whereas if m is a positive odd number, the odd series will be a polynomial. The solution which is finite at $|x| = 1$ is therefore the polynomial $R_n(x)$. In the next section we shall show that these polynomials are orthogonal and therefore multiples of the Legendre polynomials.

4.3 Orthogonality

The differential equation (4.2.1) for $R_m(x)$ can be written in Sturm Liouville form

$$\frac{d}{dx}\left((1 - x^2)\frac{dR_m(x)}{dx}\right) = -m(m+1)R_m(x). \tag{4.3.1}$$

We then multiply this by $R_n(x)$, where $n \neq m$ and integrate from -1 to 1. The left-hand side becomes

$$\int_{-1}^{1} R_n(x)\frac{d}{dx}\left((1 - x^2)R_m'(x)\right)dx.$$

On integrating this by parts and noting that the integrated term vanishes at $x = \pm 1$ we find that

$$\int_{-1}^{1} (1 - x^2) R'_n(x) R'_m(x) dx = m(m + 1) \int_{-1}^{1} R_n(x) R_m(x) dx.$$

If we now follow the same procedure as above with m and n interchanged we arrive at an equation which is the same as that above on the left-hand side but with $n(n + 1)$ on the right. If we take the difference of these equations, we get

$$[m(m + 1) - n(n + 1)] \int_{-1}^{1} R_n(x) R_m(x) dx = 0.$$

Thus for $n \neq m$,

$$\int_{-1}^{1} R_n(x) R_m(x) dx = 0. \tag{4.3.2}$$

The polynomials $R_n(x)$ thus satisfy the same orthogonality relations as the Legendre polynomials and must therefore be multiples of them. If we divide $R_n(x)$ by $R_n(1)$ so that the resulting polynomial is equal to 1 at $x = 1$ we obtain the Legendre polynomial $P_n(x)$.

4.4   Rodrigues Formula

By treating $(1 - x^2)^n$ as the product $(1 - x)^n (1 + x)^n$ and using the formula for differentiating the product uv m times, we see that if $n > m$ then the m th derivative of $(1 - x^2)^n$ contains as a factor $(1 - x^2)^{n-m}$ and so

$$\frac{d^m}{dx^m}(1 - x^2)^n = \frac{d^m}{dx^m}[(1 - x)^n(1 + x)^n] = 0 \qquad \text{when} \qquad x = \pm 1.$$

From this result, it follows that

$$\int_{-1}^{1} \frac{d^n}{dx^n}(1 - x^2)^n dx = 0.$$

If we integrate by parts we see that

$$\int_{-1}^{1} x \frac{d^n}{dx^n}(1 - x^2)^n dx = 0 \qquad \text{provided that } n > 1,$$

and integrating by parts m times

$$\int_{-1}^{1} x^m \frac{d^n}{dx^n}(1 - x^2)^n = 0 \qquad \text{provided that } n > m. \tag{4.4.1}$$

In other words, the nth order polynomial $Q_n(x) = d^n(1 - x^2)^n/dx^n$ is orthogonal to x^m for all values of $m < n$. This means that $Q_n(x)$ is orthogonal to $Q_m(x)$ for all values of $m < n$. These polynomials must therefore be multiples of the Legendre polynomials we found earlier. We can find out what this multiple is by evaluating $Q_n(1)$ by writing

$$Q_n(1) = \lim_{x \to 1} \frac{d^n}{dx^n} \left\{ (1 - x)^n (1 + x)^n \right\} = (-1)^n 2^n n!.$$

Therefore

$$P_n(x) = \frac{(-1)^n}{2^n n!} \frac{d^n}{dx^n} (1 - x^2)^n = \frac{1}{2^n n!} \frac{d^n}{dx^n} (x^2 - 1)^n. \qquad (4.4.2)$$

4.5 Explicit Expression

We can obtain an explicit expression for $P_n(x)$ as a polynomial by expanding $(x^2 - 1)^n$ and performing the differentiation:

$$P_n(x) = \frac{1}{2^n} \sum_{m=0}^{[n/2]} \frac{(-1)^m (2n - 2m)!}{m!(n - m)!(n - 2m)!} x^{n-2m}. \qquad (4.5.1)$$

The coefficient of x^n in $P_n(x)$, k_n is

$$k_n = \frac{(2n)!}{2^n (n!)^2} \qquad (4.5.2)$$

and of course $k_n' = 0$.

We can express x^n in terms of the Legendre polynomials

$$x^n = \sum_{m=0}^{n} a_m P_m(x).$$

The coefficients a_m can be calculated by multiplying this equation by $P_q(x)$ and integrating from -1 to 1. Making use of the orthogonality property of the Legendre polynomials, we see

$$a_q h_q = \int_{-1}^{1} x^n P_q(x) dx = \frac{1}{2^q q!} \int_{-1}^{1} x^n \frac{d^q}{dx^q} (x^2 - 1)^q$$

$$= \frac{n!}{2^q q!(n - q)!} \int_{-1}^{1} x^{n-q} (1 - x^2)^q dx$$

after integrating by parts q times with the integrated parts being zero at the end points. If $n - q$ is odd, this integral will be zero. Therefore let $n - q = 2m$, then

$$a_{n-2m}h_{n-2m} = \frac{n!}{2^{n-2m}(n-2m)!\,(2m)!} \int_{-1}^{1} x^{2m}(1 - x^2)^{n-2m} dx.$$

Writing $t = x^2$ converts the integral into the Beta function

$$B(m + 1/2, n - 2m + 1) = \frac{\Gamma(m + 1/2)(n - 2m)!}{\Gamma(n - m + 3/2)}.$$

(See the General Appendix for the properties of the Beta Function.) Then

$$x^n = n! \sum_{m=0}^{[n/2]} \frac{(n - 2m + 1/2)\sqrt{\pi}}{2^n\, m!\,\Gamma(n - m + 3/2)} P_{n-2m}(x), \qquad (4.5.3)$$

where $[n/2]$ is the largest integer less than or equal to $n/2$. The value of h_q from the next section has been used.

4.6 Generating Function

We can show that

$$\frac{\partial}{\partial x}\left\{(1 - x^2)\frac{\partial}{\partial x}\left[\frac{1}{\sqrt{1 - 2xt + t^2}}\right]\right\} = -t\frac{\partial^2}{\partial t^2}\left[\frac{t}{\sqrt{1 - 2xt + t^2}}\right]. \qquad (4.6.1)$$

This is most easily done using a computer algebra package such as Maple. Let us define $\phi_n(x)$ by

$$\sum_{n=0}^{\infty} t^n \phi_n(x) = \frac{1}{\sqrt{1 - 2xt + t^2}} \qquad |t| < 1. \qquad (4.6.2)$$

On expanding the right-hand side using the binomial theorem we see that the functions $\phi_n(x)$ are nth order polynomials in x. If we substitute the left hand side of Eq. (4.6.2) into Eq. (4.6.1), we see that

$$\sum_{n=0}^{\infty} t^n \frac{d}{dx}\left\{(1 - x^2)\frac{d\phi_n(x)}{dx}\right\} = -t\sum_{n=0}^{\infty} \phi_n(x)\frac{d^2}{dt^2}t^{n+1} = -\sum_{n=0}^{\infty} n(n + 1)t^n \phi_n(x). \qquad (4.6.3)$$

If we now equate the coefficients of powers of t on both sides of the equation, we obtain

$$\frac{d}{dx}\left\{(1 - x^2)\frac{d\phi_n(x)}{dx}\right\} + n(n + 1)\phi_n(x) = 0. \qquad (4.6.4)$$

In other words $\phi_n(x)$ satisfies Legendre's equation and is therefore some multiple of the Legendre polynomial $P_n(x)$.

If we put $x = 1$ in Eq. (4.6.2),

$$\sum_{n=0}^{\infty} t^n \phi_n(1) = \frac{1}{1-t} = \sum_{n=0}^{\infty} t^n.$$

Thus we see that $\phi_n(1) = 1$ and that therefore $\phi_n(x) = P_n(x)$, the nth Legendre polynomial. Thus

$$\sum_{n=0}^{\infty} t^n P_n(x) = \frac{1}{\sqrt{1 - 2xt + t^2}} \qquad |t| < 1. \tag{4.6.5}$$

We can use the generating function to derive the orthogonality relations. Consider

$$\sum_{n,m=0}^{\infty} s^m t^n \int_{-1}^{1} P_m(x) P_n(x) dx = \int_{-1}^{1} \frac{1}{\sqrt{1 - 2sx + s^2}} \frac{1}{\sqrt{1 - 2tx + t^2}} dx$$

$$= \frac{1}{\sqrt{st}} \ln\left\{\frac{1 + \sqrt{st}}{1 - \sqrt{st}}\right\} = 2 + \frac{2}{3} st + \frac{2}{5} (st)^2 + \dots = 2 \sum_{n=0}^{\infty} s^n t^n / (2n + 1).$$

[The proof of this result is algebraically a bit complicated and is given in the appendix to this chapter.]

From this we see that the coefficient of $s^m t^n$ on the left-hand side, that is $\int_{-1}^{1} P_m(x) P_n(x) dx$ is zero, confirming the orthogonality relations, and that for $n = m$,

$$h_m = \int_{-1}^{1} P_m^2(x) dx = \frac{2}{2m + 1}. \tag{4.6.6}$$

4.7   Recurrence Relations

The recurrence relation for Legendre polynomials can be found from the generating function. If we differentiate Eq. (4.6.5) with respect to t, we get

$$\sum_{n=0}^{\infty} n t^{n-1} P_n(x) = \frac{t - x}{(1 - 2xt + t^2)^{3/2}}$$

and so

$$(1 - 2xt + t^2) \sum_{n=0}^{\infty} n t^{n-1} P_n(x) = \frac{t - x}{\sqrt{1 - 2xt + t^2}} = (t - x) \sum_{n=0}^{\infty} t^n P_n(x).$$

If we equate the coefficients of t^n on both sides of this equation we obtain the recurrence relation.

$$(n + 1) P_{n+1}(x) = (2n + 1) x P_n(x) - n P_{n-1}(x). \tag{4.7.1}$$

4.8 Differential Relation

We can show that

$$(1 - x^2)\frac{\partial}{\partial x}\left(\frac{1}{\sqrt{1 - 2xt + t^2}}\right) = t\frac{\partial}{\partial t}\left(\frac{t - x}{\sqrt{1 - 2xt + t^2}}\right).$$

If we substitute the left-hand side of the generating function into the equation above and equate the coefficients of t^n on both sides of the equation we see that

$$(1 - x^2)P_n'(x) = nP_{n-1}(x) - nxP_n(x). \tag{4.8.1}$$

4.9 Step Up and Step Down Operators

These operators can be easily derived from relations (4.8.1) and (4.7.1).

$$P_{n-1}(x) = S_n^- P_n(x) = \left\{x + \frac{1}{n}(1 - x^2)\frac{d}{dx}\right\}P_n(x) \tag{4.9.1}$$

and

$$P_{n+1}(x) = S_n^+ P_n(x) = \left\{x - \frac{1}{n+1}(1 - x^2)\frac{d}{dx}\right\}P_n(x). \tag{4.9.2}$$

4.10 Appendix

$$\int_{-1}^{1} \frac{1}{\sqrt{1 - 2sx + s^2}} \frac{1}{\sqrt{1 - 2tx + t^2}} dx$$

$$= \frac{1}{2\sqrt{st}} \int_{-1}^{1} \frac{dx}{\sqrt{(s + 1/s)(t + 1/t)/4 - x(s + 1/s + t + 1/t)/2 + x^2}}$$

$$= \frac{1}{2\sqrt{st}} \int_{-1}^{1} \frac{dx}{\sqrt{(x - (s + 1/s + t + 1/t)/4)^2 - (s + 1/s - t - 1/t)^2/16}}.$$

The integral part is a standard $\cosh^{-1}$ integral

$$-\left[\cosh^{-1}\left(\frac{(s+1/s+t+1/t)/4-x}{|s+1/s-t-1/t|/4}\right)\right]^{1}_{-1}$$

$$= \ln\frac{s+1/s+t+1/t+4+\sqrt{(s+1/s+t+1/t+4)^2-(s+1/s-t-1/t)^2}}{s+1/s+t+1/t-4+\sqrt{(s+1/s+t+1/t-4)^2-(s+1/s-t-1/t)^2}}$$

$$= \ln\left(\frac{(\sqrt{s}+1/\sqrt{s})^2+(\sqrt{t}+1/\sqrt{t})^2+\sqrt{4(s+2+1/s)(t+2+1/t)}}{(\sqrt{s}-1/\sqrt{s})^2+(\sqrt{t}-1/\sqrt{t})^2+\sqrt{4(s-2+1/s)(t-2+1/t)}}\right)$$

$$= \ln\left(\frac{(\sqrt{s}+1/\sqrt{s})^2+(\sqrt{t}+1/\sqrt{t})^2+2(\sqrt{s}+1/\sqrt{s})(\sqrt{t}+1/\sqrt{t})}{(\sqrt{s}-1/\sqrt{s})^2+(\sqrt{t}-1/\sqrt{t})^2+2(\sqrt{s}-1/\sqrt{s})(\sqrt{t}-1/\sqrt{t})}\right)$$

$$= \ln\left(\frac{\sqrt{s}+1/\sqrt{s}+\sqrt{t}+1/\sqrt{t}}{1/\sqrt{s}-\sqrt{s}+1/\sqrt{t}-\sqrt{t}}\right)^2 = 2\ln\left(\frac{\sqrt{s}+1/\sqrt{s}+\sqrt{t}+1/\sqrt{t}}{1/\sqrt{s}-\sqrt{s}+1/\sqrt{t}-\sqrt{t}}\right).$$

Note that $|s| < 1$ and $|t| < 1$ so that $\sqrt{s} < 1/\sqrt{s}$.

If we denote the argument of the log by u, the log can be written in the form $\ln((1+v)/(1-v))$ if

$$u = \frac{1+v}{1-v}, \qquad \text{then} \qquad v = \frac{u-1}{u+1}$$

that is substituting

$$u = \left(\frac{\sqrt{s}+1/\sqrt{s}+\sqrt{t}+1/\sqrt{t}}{1/\sqrt{s}-\sqrt{s}+1/\sqrt{t}-\sqrt{t}}\right)$$

into the expression for v and multiplying through by the denominator to get

$$v = \frac{(\sqrt{s}+1/\sqrt{s}+\sqrt{t}+1/\sqrt{t})-(1/\sqrt{s}-\sqrt{s}+1/\sqrt{t}-\sqrt{t})}{(\sqrt{s}+1/\sqrt{s}+\sqrt{t}+1/\sqrt{t})+(1/\sqrt{s}-\sqrt{s}+1/\sqrt{t}-\sqrt{t})}$$

$$= \frac{2\sqrt{s}+2\sqrt{t}}{2/\sqrt{s}+2/\sqrt{t}} = \sqrt{st}.$$

References

Copson E T, Theory of Functions of a Complex Variable, Oxford University Press 1955.

Courant R and Hilbert D, Methods of mathematical Physics Vol 1, Interscience Publishers, 1953.

Dennery P and Krzywicki A, Mathematics for Physicists, Harper and Rowe, 1967.

Erdelyi A, Higher Transcendental Functions Vol 2, McGraw-Hill 1953.

Ferrar W L, A Text-book of Convergence, Oxford University Press, 1938.

Hochstrasser Urs W, Orthogonal Polynomials, Chapter 22, Handbook of Mathematical Functions, Eds. Abramowitz M and Stegun I A, Dover, 1970.

Koornwinder T H, Wong R, Koekoek R and Swarttouw R F, Chapter 18, NIST Handbook of Mathematical Functions, Eds. Olver W J, Lozier D W, Boisvert R F and Clark C W, NIST and Cambridge University Press, 2009.

Macrobert T M, Functions of a Complex Variable, Macmillan, 1933.

Mandl F, Quantum Mechanics, Butterworth Scientific Publications, 1957.

Phillips E G, A Course of Analysis, Cambridge University Press, 1930.

Pauling L and Wilson E B, Introduction to Quantum Mechanics, McGraw-Hill, 1935.

Schiff L I, Quantum Mechanics, McGraw-Hill, 1955.

Sneddon I N, Special Functions of Mathematical Physics and Chemistry, Oliver and Boyd, 1956.

Whittaker E T and Watson G N, A Course of Modern Analysis, Cambridge University Press, 1963.

Chapter 5

Chebyshev Polynomials of the First Kind

5.1 Introduction

Chebyshev polynomials of the first kind $T_n(x)$ can be obtained using the Gram-Schmidt orthogonalisation process for polynomials in the domain $(-1, 1)$ with the weight factor $1/\sqrt{1 - x^2}$. The resulting polynomial $R_n(x)$ is multiplied by a number which makes the value at $x = 1$ equal to 1. The resulting polynomials, $T_n(x)$ are multiples of the Gegenbauer polynomials $C_n^{(0)}(x)$ and the Jacobi polynomials $P_n^{(-1/2,-1/2)}(x)$. In fact

$$T_n(x) = nC_n^{(0)}(x)/2 = \frac{n!\sqrt{\pi}}{\Gamma(n + 1/2)} P_n^{(-1/2,-1/2)}(x). \tag{5.1.1}$$

The first few Chebyshev polynomials of the first kind are:

$T_0(x) = 1.$
$T_1(x) = x.$
$T_2(x) = 2x^2 - 1.$
$T_3(x) = 4x^3 - 3x.$
$T_4(x) = 8x^4 - 8x^2 + 1.$
$T_5(x) = 16x^5 - 20x^3 + 5x.$
$T_6(x) = 32x^6 - 48x^4 + 18x^2 - 1.$
$T_7(x) = 64x^7 - 112x^5 + 56x^3 - 7x.$
$T_8(x) = 128x^8 - 256x^6 + 160x^4 - 32x^2 + 1.$
$T_9(x) = 256x^9 - 576x^7 + 432x^5 - 120x^3 + 9x.$
$T_{10}(x) = 512x^{10} - 1280x^8 + 1120x^6 - 400x^4 + 50x^2 - 1.$

Graphs of $T_n(x)$ for $n = 1$ to $n = 6$.

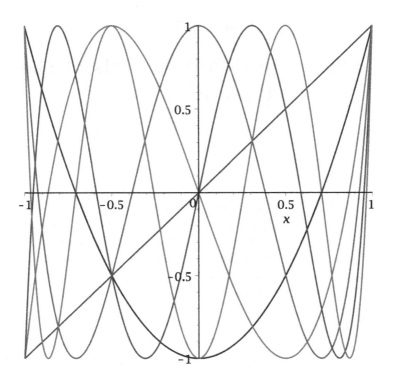

5.2 Differential Equation

Chebyshev polynomials of the first kind satisfy the ordinary diferential equation:

$$(1 - x^2)\frac{d^2y}{dx^2} - x\frac{dy}{dx} + \lambda y = 0. \qquad (5.2.1)$$

The point $x = 0$ is an ordinary point. This means that we can express the solution in the form of a power series $y = \sum_{n=0}^{\infty} a_n x^n$. The singularities of the differential equation are at $x = \pm 1$ and so the radius of convergence of the power series will be 1. On substituting this power series into the

differential equation, we find

$$(1 - x^2) \sum_{n=0}^{\infty} n(n-1)a_n x^{n-2} - x \sum_{n=0}^{\infty} n a_n x^{n-1} + \lambda \sum_{n=0}^{\infty} a_n x^n = 0.$$

If we equate the coefficients of x^n to zero we get

$$(n+2)(n+1)a_{n+2} + (\lambda - n^2)a_n = 0.$$

This leads to the recurrence relation

$$\frac{a_{n+2}}{a_n} = \frac{n^2 - \lambda}{(n+2)(n+1)}. \tag{5.2.2}$$

We see that there are two power series solutions, one series containing only even powers of x and the other an odd power series.

The ratio test can be used to show that both series converge for $|x| < 1$, but since $\lim_{n \to \infty}(a_{n+2}/a_n) = 1$, the test fails for $|x| = 1$. To investigate the convergence for $|x| = 1$ we need a more powerful test. If we let $u_m = a_{2m}$, then

$$\frac{u_m}{u_{m+1}} = 1 + \frac{1.5}{m} + O\left(\frac{1}{m^2}\right).$$

The coefficient of $1/m$ in the ratio of successive terms is 1.5 and so using Raabe's test or Gauss's test (see e.g. Philips, p. 129) the even power series converges. The odd power series also converges for $|x| \leq 1$.

Looking at the recurrence relation we see that if λ is the square of an even integer, the even power series terminates and becomes a polynomial while the odd power series remains an infinite series, and if λ is the square of an odd integer, the odd power series terminates and becomes a polynomial. Let us denote the polynomial solution with $\lambda = n^2$ by $R_n(x)$.

We are now going to show that these polynomial solutions $R_m(x)$ satisfy the orthogonality relations of the Chebyshev polynomials of the first kind and must therefore be multiples of them.

5.3   Orthogonality

The differential equation (5.2.1) for $y = R_m(x)$ can be written in Sturm Liouville form

$$\frac{d}{dx}\left(\sqrt{1 - x^2}\,\frac{dR_m(x)}{dx}\right) = -\frac{m^2 R_m(x)}{\sqrt{1 - x^2}}. \tag{5.3.1}$$

If we multiply this by $R_n(x)$, where $n \neq m$ and integrate from -1 to 1, the left-hand side becomes

$$\int_{-1}^{1} R_n(x) \frac{d}{dx} \left(\sqrt{1-x^2} R_m'(x) \right) dx.$$

On integrating this by parts and noting that the integrated term vanishes at both end points $x = \pm 1$, we find that

$$\int_{-1}^{1} \sqrt{1-x^2}\, R_n'(x) R_m'(x) dx = m^2 \int_{-1}^{1} \frac{R_n(x) R_m(x)}{\sqrt{1-x^2}} dx.$$

If we follow the same procedure as above but with m and n interchanged we will produce an equation which is the same as that above on the left-hand side and the same integral on the right but with a coefficient n^2 instead of m^2. If we subtract one of these equations from the other we get

$$[m^2 - n^2] \int_{-1}^{1} \frac{R_n(x) R_m(x)}{\sqrt{1-x^2}} dx = 0.$$

Thus for $n \neq m$

$$\int_{-1}^{1} \frac{R_n(x) R_m(x)}{\sqrt{1-x^2}} dx = 0. \tag{5.3.2}$$

The polynomials $R_n(x)$ thus satisfy the same orthogonality relations as the Chebyshev polynomials of the first kind and therefore must be multiples of them. If we divide each $R_n(x)$ by $R_n(1)$ so that the resulting polynomial is equal to 1 when $x = 1$, we will obtain the Chebyshev polynomial of the first kind.

Since x^n can be represented as a linear combination of the Chebyshev polynomials $T_p(x)$ for $0 \leq p \leq n$, we can deduce that

$$\int_{-1}^{1} \frac{x^n T_m(x) dx}{\sqrt{1-x^2}} = 0 \qquad \text{for all } n < m. \tag{5.3.3}$$

5.4 Trigonometric Representation

If we change the independent variable to θ by putting $x = \cos\theta$ so that $y(x)$ becomes $g(\theta)$, Eq. (5.2.1) with $\lambda = n^2$ becomes

$$\frac{d^2 g}{d\theta^2} + n^2 g(\theta) = 0. \tag{5.4.1}$$

The solutions of this equation are $\cos(n\theta)$ and $\sin(n\theta)$. If n is not an integer, neither solution can be expressed as a polynomial in $\cos\theta$. If n is an integer,

$\cos(n\theta)$ can be expressed as a simple polynomial in $x = \cos\theta$ but $\sin(n\theta)$ cannot. This means that the polynomial solution $R_n(x)$ of Eq. (5.2.1) is $\cos(n\theta)$ expressed as a polynomial in powers of $x = \cos\theta$. This means that, since $R_n(1) = \cos 0 = 1$, $T_n(\cos\theta) = \cos(n\theta)$. We see from this that $T_n(x)$ oscilates between $+1$ and -1.

We can prove the orthogonality relation by substituting $x = \cos\theta$ into Eq. (5.3.2) which becomes.

$$\int_0^\pi \cos(n\theta)\cos(m\theta)\,d\theta = 0 \qquad n \neq m. \tag{5.4.2}$$

The parameter h_n is

$$h_n = \int_{-1}^1 \frac{[T_n(x)]^2}{\sqrt{1-x^2}}dx = \int_0^\pi [\cos n\theta]^2 d\theta = \begin{cases} \pi/2 & n \neq 0 \\ \pi & n = 0. \end{cases} \tag{5.4.3}$$

The coefficient of x^n in $T_n(x)$ can be obtained by using Euler's formula for $\cos\theta$:

$$2^{n-1}\cos^n\theta = \left(\frac{e^{i\theta}+e^{-i\theta}}{2}\right)^n = \cos n\theta + n\cos[(n-2)\theta] + \cdots .$$

Therefore $\cos n\theta = 2^{n-1}\cos^n\theta + \cdots$. and so $k_n = 2^{n-1}$ and $k'_n = 0$.

5.5 Explicit Expression

We can obtain an explicit expression for $T_n(x)$ as a polynomial in powers of x by using the recurrence relation Eq. (5.2.2) derived in the section on the differential equation for $T_n(x)$:

$$\frac{a_{p-2}}{a_p} = \frac{p(p-1)}{(p-2-n)(p-2+n)}. \tag{5.5.1}$$

Noting that the coefficient of x^n, $k_n = 2^{n-1}$, this leads to

$$T_n(x) = \frac{n}{2}\sum_{m=0}^{[n/2]} \frac{(-1)^m(n-m-1)!}{(n-2m)!m!}(2x)^{n-2m}, \tag{5.5.2}$$

where $[n/2]$ is the largest integer less than or equal to $n/2$.

Using the binomial expansion for $(e^{i\theta} + e^{-i\theta})$, it is easy to see that

$$x^{2n+1} = 2^{-2n}\sum_{m=0}^n \frac{(2n+1)!}{m!\,(2n+1-m)!}T_{2n+1-2m}(x) \tag{5.5.3}$$

and

$$x^{2n} = 2^{1-2n}\sum_{m=0}^{n-1} \frac{(2n)!}{m!\,(2n-m)!}T_{2n-2m}(x) + 2^{-2n}\frac{(2n)!}{(n!)^2}. \tag{5.5.4}$$

5.6 Rodrigues Formula

By treating $(1-x^2)^{m-1/2}$ as the product $(1-x)^{m-1/2}(1+x)^{m-1/2}$ and using the formula to differentiate the product uv n times, we see that if $m > n$ then the nth derivative of $(1-x^2)^{m-1/2}$ contains as a factor $(1-x^2)^{m-n-1/2}$ and so

$$\frac{d^n}{dx^n}(1-x^2)^{m-1/2} = \frac{d^n}{dx^n}\left\{(1-x)^{m-1/2}(1+x)^{m-1/2}\right\} = 0 \quad \text{when } x = \pm 1.$$

From this result, it follows that

$$\int_{-1}^{1} \frac{d^n}{dx^n}(1-x^2)^{n-1/2}dx = 0.$$

If we integrate by parts we see that

$$\int_{-1}^{1} x \frac{d^n}{dx^n}(1-x^2)^{n-1/2}dx = 0 \qquad \text{provided that } n > 1,$$

and integrating by parts m times

$$\int_{-1}^{1} x^m \frac{d^n}{dx^n}(1-x^2)^{n-1/2} = 0 \qquad \text{provided that } n > m.$$

Let us define the mth order polynomial $Q_n(x)$ by

$$Q_n(x) = \sqrt{1-x^2}\,\frac{d^n}{dx^n}\left\{\frac{(1-x^2)^n}{\sqrt{1-x^2}}\right\}. \tag{5.6.1}$$

Then

$$\int_{-1}^{1} \frac{x^m Q_n(x)}{\sqrt{1-x^2}}dx = 0 \qquad m < n.$$

In other words, the nth order polynomial $Q_n(x)$ is orthogonal to x^m for all values of $m < n$. This means that $Q_m(x)$ is orthogonal to $Q_n(x)$ for all values of $m < n$. These polynomials must therefore be multiples of the Chebyshev polynomials we found earlier. We can find the multiplication constant by evaluating $Q_n(1)$.

$$Q_n(1) = \lim_{x \to 1} \sqrt{1-x^2}\,\frac{d^n}{dx^n}\left\{(1-x)^{n-1/2}(1+x)^{n-1/2}\right\}.$$

The only term which contributes in the limit as $x \to 1$ is $(1+x)^n\sqrt{1-x}$ multiplied by the nth derivative of $(1-x)^{n-1/2}$.

$$Q_n(1) = (n-1/2)(n-3/2)...(1/2)2^n(-1)^n.$$

Thus, since $T_n(1) = 1$,

$$T_n(x) = \frac{(-1)^n}{1.3.5.7...(2n-1)}\sqrt{1-x^2}\,\frac{d^n}{dx^n}\left\{\frac{(1-x^2)^n}{\sqrt{1-x^2}}\right\}. \tag{5.6.2}$$

5.7 Generating Functions

The simplest generating function can be derived by taking the real part of the identity

$$\sum_{n=0}^{\infty} t^n e^{in\theta} = \frac{1}{1 - te^{i\theta}}. \tag{5.7.1}$$

That is

$$\sum_{n=0}^{\infty} t^n \cos n\theta = Re\left(\frac{1}{1 - te^{i\theta}}\right).$$

Thus putting $x = \cos\theta$. and so

$$\sum_{n=0}^{\infty} t^n T_n(x) = \frac{1 - xt}{1 - 2xt + t^2}. \tag{5.7.2}$$

The derivation of the generating function (5.7.2) arises as a special case of using the identity

$$\sum_{n=0}^{\infty} \frac{\Gamma(n+\mu)}{n!\,\Gamma(\mu)}(te^{i\theta})^n = 1 + \mu te^{i\theta} + \frac{\mu(\mu+1)}{1.2}t^2 e^{2i\theta} + \dots = \frac{1}{(1 - te^{i\theta})^\mu} \tag{5.7.3}$$

and taking the real part on both sides.

If we let $\mu = 2$ we can derive another particular generating function

$$\sum_{n=0}^{\infty}(n+1)t^n T_n(x) = \frac{1 - 2tx + 2t^2 x^2 - t^2}{(1 - 2tx + t^2)^2}. \tag{5.7.4}$$

Another example can be derived from (5.7.3) by taking $\mu = 1/2$. We start by writing (5.7.3) in the form

$$\sum_{n=0}^{\infty} \frac{\Gamma(n+1/2)}{n!\,\Gamma(1/2)}(te^{i\theta})^n = \sum_{n=0}^{\infty} \frac{(2n)!}{(n!)^2}\frac{t^n e^{ni\theta}}{4^n}$$

$$= 1 + \frac{1}{2}te^{i\theta} + \frac{\frac{1}{2}\cdot\frac{3}{2}}{1.2}t^2 e^{2i\theta} + \frac{\frac{1}{2}\cdot\frac{3}{2}\cdot\frac{5}{2}}{1.2.3}t^3 e^{3i\theta}. + \dots = \frac{1}{\sqrt{1 - te^{i\theta}}}.$$

We now take the real part on both sides of the above equation. Thus

$$\sum_{n=0}^{\infty} \frac{(2n)!}{(n!)^2}\frac{t^n T_n(x)}{4^n} = \frac{\sqrt{1 - tx + \sqrt{1 - 2tx + t^2}}}{\sqrt{2}\sqrt{1 - 2tx + t^2}}. \tag{5.7.5}$$

This generating function is a special case of that for the Jacobi polynomials $P_n^{(-1/2,-1/2)}(x)$, noting that $T_n(x) = [n!\sqrt{\pi}/\Gamma(n+1/2)]P_n^{(-1/2,-1/2)}(x)$.

A further example can be derived by putting $\mu = -1/2$ in (5.7.3) and using the binomial expansion of

$$\sum_{n=0}^{\infty} \frac{\Gamma(n-1/2)}{\Gamma(-1/2)\, n!} t^n e^{ni\theta} = 1 - \frac{1}{2} t e^{i\theta} - \frac{\frac{1}{2} \cdot \frac{1}{2}}{2} t^2 e^{2i\theta} + \ldots = \sqrt{1 - t e^{i\theta}}.$$

On taking the real part on both sides we find

$$\sum_{n=0}^{\infty} \sqrt{2} \frac{\Gamma(n-1/2)}{\Gamma(-1/2)\, n!} t^n T_n(x) = \sqrt{1 - tx + \sqrt{1 - 2tx + t^2}}. \qquad (5.7.6)$$

A further generating function can be derived from the identity

$$\sum_{n=1}^{\infty} \frac{t^n e^{in\theta}}{n} = -\ln(1 - t e^{i\theta}).$$

On taking the real part and writing $x = \cos\theta$, we find

$$\sum_{n=1}^{\infty} \frac{t^n}{n} T_n(x) = -Re\{\ln(1 - e^{i\theta})\} = -\frac{1}{2} \ln(1 - 2tx + t^2). \qquad (5.7.7)$$

An alternative demonstration of the validity of these generating functions is to define functions $\phi_n(x)$ by

$$\sum_{n=0}^{\infty} t^n \phi_n(x) = w(x, t) \qquad (5.7.8)$$

for each function $w(x, t)$ on the right-hand side of (5.7.2) and (5.7.4) - (5.7.7). We can see in each case that $\phi_n(x)$ is an n th order polynomial in x by using the binomial theorem to expand the appropriate $w(x, t)$ in powers of t and collecting the terms in t^n. The next step is to show that for each possible function w

$$(1 - x^2) \frac{\partial^2 w(x, t)}{\partial x^2} - x \frac{\partial w(x, t)}{\partial x} = -t \frac{\partial}{\partial t} \left(t \frac{\partial w(x, t)}{\partial t} \right). \qquad (5.7.9)$$

This can be done using a computer algebra package such as Maple.

If we substitute the left-hand side of (5.7.8) into (5.7.9) we see that

$$\sum_{n=0}^{\infty} t^n \left\{ (1 - x^2) \frac{d^2 \phi_n(x)}{dx^2} - x \frac{d\phi_n(x)}{dx} \right\} = -t \sum_{n=0}^{\infty} \phi_n(x) \frac{d}{dt} \left\{ t \frac{d}{dt} t^n \right\}$$

$$= -\sum_{n=0}^{\infty} n^2 t^n \phi_n(x). \qquad (5.7.10)$$

If we now equate the coefficients of powers of t on both sides of the equation we find

$$(1 - x^2)\frac{d^2\phi_n(x)}{dx^2} - x\frac{d\phi_n(x)}{dx} + n^2\phi_n(x) = 0. \tag{5.7.11}$$

In other words, $\phi_n(x)$ satisfies Chebyshev's equation and therefore must be multiple of the Chebyshev polynomial of the first kind $T_n(x)$.

If we put $x = 1$ in $w(x, t)$ for each function on the right-hand side of Eqs. (5.7.2), (5.7.4), (5.7.5) and (5.7.6) and expand in powers of t, we can find the factor a_n such that $T_n(x) = a_n\phi_n(x)$.

For (5.7.2), $w(x, t) = (1 - xt)/(1 - 2xt + t^2)$, $w(1, t) = 1/(1 - t)$,

$$\sum_{n=0}^{\infty} t^n \phi_n(1) = \frac{1}{1 - t} = \sum_{n=0}^{\infty} t^n,$$

so that $\phi_n(1) = 1$, $a_n = 1$ and therefore $\phi_n(x) = T_n(x)$ confirming the form of the left-hand side of Eq. (5.7.2).

For the generating function (5.7.3),

$$w(1, t) = 1/(1 - t)^2 = 1 + 2t + 3t^2 + \ldots = \sum_{n=0}^{\infty} (n + 1)t^n$$

and so in this case $\phi_n(x) = (n + 1)T_n(x)$.

For the generating function (5.7.5),

$$w(1, t) = \frac{1}{\sqrt{1 - t}} = 1 + \frac{1}{2}t + \frac{\frac{1}{2}\cdot\frac{3}{2}}{2}t^2 = \sum_{n=0}^{\infty} \frac{\Gamma(n + 1/2)}{\Gamma(1/2)n!}t^n$$

and so in this case,

$$\phi_n(x) = \frac{\Gamma(n + 1/2)}{n!\,\Gamma(1/2)}T_n(x)$$

which is in agreement with the left-hand side of (5.7.5).

For the generating function (5.7.6),

$$w(1, t) = \sqrt{2}\sqrt{1 - t} = \sqrt{2}\left(1 - \frac{1}{2}t - \frac{\frac{1}{2}\cdot\frac{1}{2}}{2}t^2 + \ldots\right) = \sum_{n=0}^{\infty} \sqrt{2}\frac{\Gamma(n - 1/2)}{n!\,\Gamma(-1/2)}t^n$$

and so in this case,

$$\phi_n(x) = \sqrt{2}\frac{\Gamma(n - 1/2)}{n!\,\Gamma(-1/2)}T_n(x).$$

For the generating function (5.7.7),

$$w(1, t) = -\ln(1 - t) = t + \frac{t^2}{2} + \frac{t^3}{3} + \ldots$$

and so in this case,

$$\phi_n(x) = \frac{T_n(x)}{n}.$$

5.8 Recurrence Relations

There is a simple recurrence relation between the Chebyshev polynomials which is most easily derived using the trigonometric identity
$\cos(A + B) = \cos A \cos B - \sin A \sin B$

$$T_{n+1}(x) + T_{n-1}(x) = 2xT_n(x). \tag{5.8.1}$$

A more general recurrence relation is

$$T_{n+m}(x) + T_{n-m}(x) = 2T_n(x)T_m(x). \tag{5.8.2}$$

5.9 Addition Formulae

The generating function can be used to show that

$$\sum_{m=0}^{n} T_m(x)x^{n-m} = U_n(x), \tag{5.9.1}$$

where $U_m(x)$ is the Chebyshev polynomial of the second kind.

Multiplying Eq. (5.9.1) by t^n, summing from 0 to ∞ and then changing the order of the summations gives

$$\sum_{n=0}^{\infty} t^n \sum_{m=0}^{n} T_m(x)x^{n-m} = \sum_{m=0}^{\infty} T_m(x) \sum_{n=m}^{\infty} t^n x^{n-m}$$

$$= \frac{1}{1 - xt} \sum_{m=0}^{\infty} t^m T_m(x) = \frac{1}{1 - 2xt + t^2} = \sum_{m=0}^{\infty} t^m U_m(x),$$

on using the generating function Eq. (5.7.2). Equating the coefficients of t^m on both sides leads to Eq. (5.9.1).

The identity $\sin[(n + 3)\theta] - \sin[(n + 1)\theta) = 2\cos[(n + 2)\theta]\sin\theta$ leads to the identity

$$U_n(x) - U_{n-2}(x) = 2T_n(x), \tag{5.9.2}$$

and if this is iterated

$$2 \sum_{m=0}^{n} T_{2m}(x) = 1 + U_{2n}(x) \tag{5.9.3}$$

and

$$2 \sum_{m=0}^{n} T_{2m+1}(x) = U_{2n+1}(x). \tag{5.9.4}$$

The identity $\cos(n\theta) - \cos[(n-2)\theta] = 2\sin[(n+1)\theta]\sin\theta$ leads to

$$T_n(x) - T_{n-2}(x) = 2(1-x^2)U_n(x). \tag{5.9.5}$$

If this is iterated we find

$$2(1-x^2)\sum_{m=0}^{n} U_{2m} = 1 - T_{2n+2}(x) \tag{5.9.6}$$

and

$$2(1-x^2)\sum_{m=0}^{n} U_{2m+1}(x) = x - T_{2n+3}(x). \tag{5.9.7}$$

5.10   Differential Relations

We can deduce a relation for the derivative $T_n'(x)$ most easily by substituting $x = \cos\theta$. We find

$$\frac{d}{dx}T_n(x) = nU_{n-1}(x), \tag{5.10.1}$$

where $U_n(x)$ is the Chebyshev polynomial of the second kind.

This can be combined with the relation Eq. (5.9.5) to produce

$$(1-x^2)T_n'(x) = \frac{n}{2}\Big(T_{n-1}(x) - T_{n+1}(x)\Big). \tag{5.10.2}$$

5.11   Relations with Other Chebyshev Polynomials

These relations can all be proved using the trigonometric representation.

The relation $\sin[(n+1)\theta] - \sin[(n-1)\theta] = 2\cos(n\theta)\sin\theta$ leads to

$$2T_n(x) = U_n(x) - U_{n-2}(x). \tag{5.11.1}$$

The relation $\cos(n\theta) - \cos[(n+2)\theta] = 2\sin[(n+1)\theta]\sin\theta$ leads to

$$T_n(x) - T_{n+2}(x) = 2(1-x^2)U_n(x). \tag{5.11.2}$$

The relation $\sin[(n+1/2)\theta] - \sin[(n-1/2)\theta] = 2\cos(n\theta)\sin(\theta/2)$ leads to

$$2T_n(x) = W_n(x) - W_{n-1}(x). \tag{5.11.3}$$

The relation $\cos[(n+1)\theta] + \cos(n\theta) = 2\cos[(n+1/2)\theta]\cos(\theta/2)$ leads to

$$T_{n+1}(x) + T_n(x) = (1+x)V_n(x) \tag{5.11.4}$$

and similarly $\cos[(n+1)\theta] - \cos(n\theta) = 2\sin[(n+1/2)\theta]\sin(\theta/2)$ leads to

$$T_n(x) - T_{n+1}(x) = (1-x)W_n(x). \tag{5.11.5}$$

5.12 Step Up and Step Down Operators

These operators can be easily derived from relations (5.8.1) and (5.10.3)

$$T_{n-1}(x) = S_n^- T_n(x) = \left\{ x + \frac{1}{n}(1 - x^2)\frac{d}{dx} \right\} T_n(x) \qquad (5.12.1)$$

and

$$T_{n+1}(x) = S_n^+ T_n(x) = \left\{ x - \frac{1}{n}(1 - x^2)\frac{d}{dx} \right\} T_n(x). \qquad (5.12.2)$$

References

Doman B G S, International Journal of Pure and Applied Mathematics, 2010, 63, pp. 197-205.

Ferrar W L, Textbook of Convergence, Oxford University Press, 1938.

Hochstrasser Urs W, Orthogonal Polynomials, Chapter 22, Handbook of Mathematical Functions, Eds. Abramowitz M and Stegun I A, Dover, 1970.

Koornwinder T H, Wong R, Koekoek R and Swarttouw R F, Chapter 18, NIST Handbook of Mathematical Functions, Eds. Olver W J, Lozier D W, Boisvert R F and Clark C W, NIST and Cambridge University Press, 2009.

Mason J C and Handscomb D C, Chebyshev Polynomials, Chapman & Hall, 2002.

Phillips E G, A Course of Analysis, Cambridge University Press, 1930.

Chapter 6

Chebyshev Polynomials of the Second Kind

6.1 Introduction

Chebyshev polynomials of the second kind $U_n(x)$ can be obtained using the Gram-Schmidt orthogonalisation process for polynomials in the domain $(-1, 1)$ with the weight factor $\sqrt{1 - x^2}$. The resulting polynomial $R_n(x)$ is multiplied by a number which makes the value at $x = 1$ equal to $n+1$. The resulting polynomials, $U_n(x)$, are multiples of the Gegenbauer polynomials $C_n^{(1)}(x)$ and also the Jacobi polynomials $P_n^{(1/2,1/2)}(x)$. In fact

$$U_n(x) = C_n^{(1)}(x) = \frac{(n + 1)!\Gamma(3/2)}{\Gamma(n + 3/2)} P_n^{(1/2,1/2)}(x). \qquad (6.1.1)$$

The first few Chebyshev polynomials of the second kind are:
$U_0(x) = 1.$
$U_1(x) = 2x.$
$U_2(x) = 4x^2 - 1.$
$U_3(x) = 8x^3 - 4x.$
$U_4(x) = 16x^4 - 12x^2 + 1.$
$U_5(x) = 32x^5 - 32x^3 + 6x.$
$U_6(x) = 64x^6 - 80x^4 + 24x^2 - 1.$
$U_7(x) = 128x^7 - 192x^5 + 80x^3 - 8x.$
$U_8(x) = 256x^8 - 448x^6 + 240x^4 - 40x^2 + 1.$
$U_9(x) = 512x^9 - 1024x^7 + 672x^5 - 160x^3 + 10x.$
$U_{10}(x) = 1024x^{10} - 2304x^8 + 1792x^6 - 560x^4 + 60x^2 + 1.$

Graphs of $U_n(x)$ for $n = 1$ to $n = 5$.

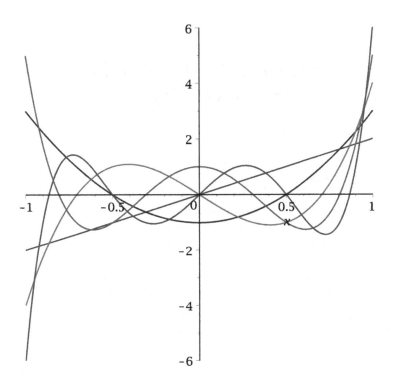

6.2 Differential Equation

The ordinary differential equation satisfied by Chebyshev polynomials of the second kind is

$$(1 - x^2)\frac{d^2 y}{dx^2} - 3x\frac{dy}{dx} + \lambda y = 0. \tag{6.2.1}$$

The point $x = 0$ is an ordinary point. This means that we can express the solution in the form of a power series $y = \sum_{n=0}^{\infty} a_n x^n$. The singularities of the differential equation are at $x = \pm 1$ and so the radius of convergence of the power series will be 1. On substituting this power series into the

differential equation, we find

$$(1 - x^2) \sum_{n=0}^{\infty} n(n-1) a_n x^{n-2} - 3x \sum_{n=0}^{\infty} n a_n x^{n-1} + \lambda \sum_{n=0}^{\infty} a_n x^n = 0.$$

If we equate the coefficients of x^n to zero we get

$$(n+2)(n+1) a_{n+2} + (\lambda - n(n+2)) a_n = 0.$$

This leads to the recurrence relation

$$\frac{a_{n+2}}{a_n} = \frac{n(n+2) - \lambda}{(n+2)(n+1)}. \tag{6.2.2}$$

We see that there are two power series solutions, one series containing only even powers of x and the other an odd power series.

The ratio test can be used to show that both series converge for $|x| < 1$, but since $\lim_{n \to \infty} (a_{n+2}/a_n) = 1$, the test fails for $|x| = 1$. To investigate the convergence for $|x| = 1$, we need a more powerful test. For the even series, if we let $u_m = a_{2m}$, then

$$\frac{u_m}{u_{m+1}} = 1 + \frac{0.5}{m} + O\left(\frac{1}{m^2}\right). \tag{6.2.3}$$

The coefficient of $1/m$ is $0.5 < 1$ and so by Raabe's test or Gauss's test, (see e.g. Phillips, p. 129) the series diverges. We can use the same argument to show that the odd series also diverges for $|x| = 1$.

Looking at the recurrence relation we see that if $\lambda = n(n+2)$ where n is a positive integer, one of the power series terminates. If this n is an even number, the even power series becomes an nth order polynomial, while if this n is odd, the odd series becomes a polynomial. Let us denote the polynomial solution by $R_n(x)$.

We are now going to show that $R_n(x)$ satisfy the orthogonality relations for Chebyshev polynomials of the second kind and must therefore be multiples of them.

6.3 Orthogonality

The differential equation (6.2.1) for $R_m(x)$ can be written in Sturm Liouville form

$$\frac{d}{dx}\left((1-x^2)^{3/2} \frac{dR_m(x)}{dx}\right) = -m(m+2)\sqrt{1-x^2}\, R_m(x). \tag{6.3.1}$$

If we multiply this by $R_n(x)$, where $n \neq m$ and integrate from -1 to 1 the left-hand side becomes

$$\int_{-1}^{1} R_n(x) \frac{d}{dx} \left((1-x^2)^{3/2} R'_m(x) \right) dx.$$

On integrating by parts and noting that the integrated term vanishes at both end points $x = \pm 1$, we find that

$$\int_{-1}^{1} (1-x^2)^{3/2} R'_n(x) R'_m(x) dx = m(m+2) \int_{-1}^{1} \sqrt{1-x^2} \, R_n(x) R_m(x) dx.$$

If we follow the same procedure as above but with m and n interchanged, we will produce an equation which is the same as that above on the left and the same integral on the right but with a coefficient $n(n+2)$ instead of $m(m+2)$. If we subtract one of these equations from the other, we get

$$[m(m+2) - n(n+2)] \int_{-1}^{1} \sqrt{1-x^2} \, R_n(x) R_m(x) = 0,$$

and so for $n \neq m$

$$\int_{-1}^{1} \sqrt{1-x^2} \, R_n(x) R_m(x) dx = 0. \qquad (6.3.2)$$

The polynomials $R_n(x)$ thus satify the same orthogonality relations as the Chebyshev polynomials of the second kind and must therefore be multiples of them. If we multiply $R_n(x)$ by $(n+1)/R_n(1)$ the resulting polynomial will equal $n+1$ when $x = 1$ and so will be the Chebyshev polynomial of the second kind.

Since x^n can be represented as a linear combination of Chebyshev polynomials of the second kind $U_p(x)$ for $0 \leq p \leq n$, we can deduce that

$$\int_{-1}^{1} \sqrt{1-x^2} \, x^n U_m(x) dx = 0 \qquad \text{for all } n < m. \qquad (6.3.3)$$

6.4   Trigonometric Representation

If we change to θ as the independent variable by putting $x = \cos\theta$ and let $g(\theta) = \sin\theta \, y(x)$ in the differential equation Eq. (6.2.1) and let $\lambda = n(n+2)$ we obtain

$$\frac{d^2 g}{d\theta^2} + (n+1)^2 g(\theta) = 0. \qquad (6.4.1)$$

The solutions of this equation are $\cos[(n+1)\theta]$ and $\sin[(n+1)\theta]$. The solution of Eq. (6.2.1), $y = \cos[(n+1)\theta]/\sin\theta$ is singular at $\theta = 0$ and therefore cannot be represented by a polynomial in $\cos\theta$. This will not be a satisfactory solution to our problem. The acceptable polynomial solution is then $R_n(x) = \sin[(n+1)\theta]/\sin\theta$, where $x = \cos\theta$. Since $x \to 1$ as $\theta \to 0$, $R_n(1) = n+1$ and therefore

$$U_n(\cos\theta) = \frac{\sin[(n+1)\theta]}{\sin\theta}. \tag{6.4.2}$$

The orthogonality relation can be proved by substituting $x = \cos\theta$ and using the trigonometric representation $U_n(\cos\theta) = \sin[(n+1)\theta]/\sin\theta$ in Eq. (6.3.2) which gives

$$\int_0^\pi \sin[(n+1)\theta)]\sin[(m+1)\theta)]d\theta = 0 \qquad n \neq m. \tag{6.4.3}$$

The parameter h_n is given by

$$h_n = \int_{-1}^1 \sqrt{1-x^2}\,[U_n(x)]^2 dx = \int_0^\pi [\sin(n+1)\theta]^2 d\theta = \frac{\pi}{2}. \tag{6.4.4}$$

The coefficient of x^n in $U_n(x)$ can be obtained by using the Euler formula for $\sin\theta$ and

$$e^{i(n+1)\theta} - e^{-i(n+1)\theta} = (e^{i\theta} - e^{-i\theta})(e^{in\theta} + e^{i(n-2)\theta} + \ldots e^{-i(n-2)\theta} + e^{-in\theta}).$$

Therefore

$$\frac{\sin(n+1)\theta}{\sin\theta} = 2\cos n\theta + 2\cos[(n-2)\theta] + \ldots$$

and

$$2^{n-1}[\cos\theta]^n = \cos n\theta + n\cos(n-2)\theta + \ldots \quad .$$

Thus

$$\frac{\sin(n+1)\theta}{\sin\theta} = 2^n[\cos\theta]^n + O(\cos\theta)^{n-2}.$$

Therefore the coefficient of x^n in $U_n(x)$, $k_n = 2^n$ and $k_n' = 0$.

6.5 Explicit Expression

We can obtain an explicit expression for $U_n(x)$ as a polynomial in powers of x by using the recurrence relation derived in the section on the differential equation for $U_n(x)$:

$$\frac{a_{p-2}}{a_p} = \frac{p(p-1)}{(p-2-n)(p+n)}. \tag{6.5.1}$$

On taking the coefficient of $x^n = 2^n$, this leads to

$$U_n(x) = \sum_{m=0}^{[n/2]} \frac{(-1)^m (n-m)!}{m!(n-2m)!} (2x)^{n-2m}, \tag{6.5.2}$$

where $[n/2]$ is the largest integer less than or equal to $n/2$.

The same result can be obtained by expanding the generating function (6.7.2) below and taking the coefficient of t^n.

We can express x^n in terms of the Chebyshev polynomials of the second kind

$$x^n = \sum_{m=0}^{n} a_m U_m(x).$$

The coefficients a_m can be calculated by multiplying this equation by $U_q(x)$ and integrating from -1 to 1. Making use of the orthogonality property of the Chebyshev polynomials and using the Rodrigues formula from the next section, we see that

$$a_q h_q = \int_{-1}^{1} x^n U_q(x) dx = \frac{(-2)^q (q+1)!}{(2q+1)! \, q!} \int_{-1}^{1} x^n \frac{d^q}{dx^q} \left[(1-x^2)^q \sqrt{1-x^2} \right] dx$$

$$= \frac{2^q (q+1)! \, n!}{(2q+1)!(n-q)!} \int_{-1}^{1} x^{n-q} (1-x^2)^q \sqrt{1-x^2} \, dx$$

after integrating by parts q times with the integrated parts being zero at the end points. If $n-q$ is odd, this integral will be zero. Therefore let $n-q = 2m$, then

$$a_{n-2m} h_{n-2m} = \frac{2^{n-2m} \, n! \, (n-2m+1)!}{(2n-4m+1)! \, (2m)!} \int_{-1}^{1} x^{2m} (1-x^2)^{n-2m+1/2} dx.$$

Writing $t = x^2$ converts the integral into the Beta function

$$B(m+1/2, n-2m+3/2) = \frac{\Gamma(m+1/2)\Gamma(n-2m+3/2)!}{\Gamma(n-m+2)}.$$

(See the General Appendix for the properties of the Beta function.) Then

$$x^n = \sum_{m=0}^{[n/2]} \frac{n! \, (n-2m+1)}{2^n \, m! \, (n-m+1)!} U_{n-2m}(x), \tag{6.5.3}$$

where $[n/2]$ is the largest integer less than or equal to $n/2$. The value of h_q from the previous section has been used.

This result can also be obtained using the trigonometric representation of $U_n(x)$. Noting that

$$\left(e^{i\theta} + e^{-i\theta}\right)^n (e^{i\theta} - e^{i\theta}) = \sum_{m=0}^{n} \frac{n!}{m!\,(n-m)!} \left\{ e^{(n+1-2m)\theta} - e^{(n-1-2m)\theta} \right\}$$

$$= 2i\sin\left[(n+1)\theta\right] + \sum_{m=1}^{n} \left\{ \frac{n!}{m!\,(n-m)!} - \frac{n!}{(m-1)!\,(n+1-m)!} \right\} e^{(n+1-2m)i\theta}.$$

In the sum, we combine terms in m and $n+1-m$ and note that if n is odd there is no term for $[n/2]+1$ to get

$$2^n \cos^n\theta \sin\theta = \sum_{m=0}^{[n/2]} \frac{n!\,(n-2m+1)}{m!\,(n-m+1)!} \sin[(n+1-2m)\theta],$$

which is the same as Eq. (6.5.3).

6.6 Rodrigues Formula

By treating $(1-x^2)^{m+1/2}$ as the product $(1-x)^{m+1/2}(1+x)^{m+1/2}$ and using the formula for differentiating the product uv n times we see that if $m > n$ then the nth derivative of $(1-x^2)^{m+1/2}$ contains as a factor $(1-x^2)^{m-n+1/2}$ and so

$$\frac{d^n}{dx^n}(1-x^2)^{m+1/2} = \frac{d^n}{dx^n}(1-x)^{m+1/2}(1+x)^{m+1/2} = 0 \qquad \text{when } x = \pm 1.$$

From this result, it follows that

$$\int_{-1}^{1} \frac{d^n}{dx^n}(1-x^2)^{n+1/2}dx = 0.$$

If we integrate by parts we see that

$$\int_{-1}^{1} x\frac{d^n}{dx^n}(1-x^2)^{n+1/2}dx = 0 \qquad \text{provided that } n > 1$$

and integrating by parts m times

$$\int_{-1}^{1} x^m \frac{d^n}{dx^n}(1-x^2)^{n+1/2} = 0 \qquad \text{provided that } n > m.$$

Let us define the nth order polynomial $Q_n(x)$ by

$$Q_n(x) = \frac{1}{\sqrt{1-x^2}} \frac{d^n}{dx^n}\left\{ \sqrt{1-x^2}\,(1-x^2)^n \right\}. \qquad (6.6.1)$$

Then

$$\int_{-1}^{1} \sqrt{1-x^2}\, x^m Q_n(x)dx = 0 \qquad m < n. \tag{6.6.2}$$

In other words, the nth order polynomial $Q_n(x)$ is orthogonal to x^m for all values of $m < n$. This means that $Q_m(x)$ is orthogonal to $Q_n(x)$ for all values of $m < n$. These polynomials satisfy the orthogonality relations for the Chebyshev polynomials of the second kind and must therefore be multiples of them.

We can find out what this multiple is by evaluating $Q_n(1)$.

$$Q_n(1) = \lim_{x \to 1} \frac{1}{\sqrt{1-x^2}} \frac{d^n}{dx^n} \left\{ (1-x)^{n+1/2}(1+x)^{n+1/2} \right\}.$$

The only term which contributes in the limit as $x \to 1$ is $(1+x)^{n+1/2}$ multiplied by the nth derivative of $(1-x)^{n+1/2}$.

Thus

$$Q_n(1) = (n+1/2)(n-1/2)...(3/2)2^n(-1)^n,$$

and hence

$$U_n(x) = \frac{(-1)^n(n+1)}{1.3.5.7...(2n+1)} \frac{1}{\sqrt{1-x^2}} \frac{d^n}{dx^n} \left\{ (1-x^2)^n \sqrt{1-x^2} \right\}. \tag{6.6.3}$$

6.7 Generating Functions

The simplest generating function can be derived from the identity

$$\sum_{n=0}^{\infty} t^n e^{i(n+1)\theta} = \frac{e^{i\theta}}{1 - te^{i\theta}} \qquad |t| < 1. \tag{6.7.1}$$

Taking the imaginary part of this equation gives

$$\sum_{n=0}^{\infty} t^n \sin[(n+1)\theta] = \frac{\sin\theta}{1 - 2t\cos\theta + t^2}.$$

If we divide by $\sin\theta$ and put $x = \cos\theta$ we obtain

$$\sum_{n=0}^{\infty} t^n U_n(x) = \frac{1}{1 - 2tx + t^2}. \tag{6.7.2}$$

The derivation of this generating function is a special case of using the identity

$$\sum_{n=0}^{\infty} \frac{\Gamma(n+\mu)}{n!\,\Gamma(\mu)} t^n e^{in\theta} = 1 + \mu t e^{i\theta} + \frac{\mu(\mu+1)}{1\cdot 2} t^2 e^{2i\theta} + ... = \frac{1}{(1 - te^{i\theta})^\mu}. \tag{6.7.3}$$

If we let $\mu = 2$, we have

$$\sum_{n=0}^{\infty}(n+1)t^n e^{i(n+1)\theta} = \frac{e^{i\theta}}{(1-te^{i\theta})^2}.$$

On taking the imaginary part on both sides, dividing through by $\sin\theta$ and putting $x = \cos\theta$ we find

$$\sum_{n=0}^{\infty}(n+1)t^n U_n(x) = \frac{1-t^2}{(1-2tx+t^2)^2}. \tag{6.7.4}$$

If we add Eq. (6.7.2) and Eq. (6.7.4) we get

$$\sum_{n=0}^{\infty}(n+2)U_n(x) = 2\frac{1-xt}{(1-2xt+t^2)^2}. \tag{6.7.5}$$

We get another generating function by letting $\mu = 1/2$ in Eq. (6.7.3). Specifically

$$\sum_{n=0}^{\infty}\frac{\Gamma(n+1/2)}{n!\,\Gamma(1/2)}t^n e^{in\theta} = \frac{1}{\sqrt{1-te^{i\theta}}} = \frac{1}{\sqrt{2R}\sqrt{R+a}}\left\{R+a+it\sin\theta\right\},$$

where $R = \sqrt{1-2xt+t^2}$ and $a = 1-xt$.

We now take the imaginary part on both sides of this equation

$$\sum_{n=0}^{\infty}\frac{\Gamma(n+1/2)}{n!\,\Gamma(1/2)}t^n \sin n\theta = \sum_{n=0}^{\infty}\frac{\Gamma(n+3/2)}{(n+1)!\,\Gamma(1/2)}t^{(n+1)} \sin[(n+1)\theta]$$

$$= \frac{it\sin\theta}{\sqrt{2}\,R\sqrt{R+a}},$$

where the term for $n=0$ in the first sum does not contribute. If we divide through by $t\sin\theta$, we get

$$\sum_{n=0}^{\infty}\frac{\Gamma(n+3/2)}{(n+1)!\,\Gamma(1/2)}t^n U_n(x) = \frac{1}{\sqrt{2(1-2xt+t^2)}\sqrt{\sqrt{1-2xt+t^2}+1-xt}}. \tag{6.7.6}$$

We get a further generating function by letting $\mu = -1/2$ in Eq. (6.7.3). This gives

$$\sum_{n=0}^{\infty}\frac{\Gamma(n-1/2)}{n!\,\Gamma(-1/2)}t^n e^{in\theta} = \sqrt{1-te^{i\theta}} = \frac{1}{\sqrt{2}}\left\{\sqrt{R+a} - \frac{it\sin\theta}{\sqrt{R+a}}\right\}.$$

On taking the imaginary part, dividing through by $t\sin\theta$, and again noting that the term for $n=0$ in the above equation does not contribute, we get

$$\sum_{n=0}^{\infty}\frac{\Gamma(n+1/2)}{(n+1)!\,\Gamma(-1/2)}t^n U_n(x) = \frac{-1}{\sqrt{2(\sqrt{1-2xt+t^2}+1-xt)}}. \tag{6.7.7}$$

We can obtain a more complicated generating function by integrating the generating function (6.7.2) with respect to t. After some manipulation, this gives

$$\sum_{n=0}^{\infty} \frac{t^{n+1} U_n(x)}{n+1} = \frac{1}{\sqrt{1-x^2}} \tan^{-1}\left\{ \frac{t\sqrt{1-x^2}}{1-xt} \right\}. \tag{6.7.8}$$

An alternative demonstration of the validity of these generating functions is to define functions $\phi_n(x)$ by

$$\sum_{n=0}^{\infty} t^n \phi_n(x) = w(x,t), \tag{6.7.9}$$

for each function $w(x,t)$ on the right-hand side of (6.7.2) and (6.7.4) - (6.7.8). We can see that in each case $\phi_n(x)$ is an nth order polynomial in x by using the binomial theorem to expand the appropriate $w(x,t)$ in powers of t and collecting the terms in t^n.

The next step is to show that for each $w(x,t)$,

$$(1-x^2)\frac{\partial^2 w(x,t)}{\partial x^2} - 3x\frac{\partial w(x,t)}{\partial x} = -t\frac{\partial}{\partial t}\left[\frac{1}{t}\frac{\partial t^2 w(x,t)}{\partial t} \right]. \tag{6.7.10}$$

This can conveniently be done in each case using a computer algebra package such as Maple. We then substitute the left-hand side of Eq. (6.7.9) into Eq. (6.7.10).

$$\sum_{n=0}^{\infty} t^n \left\{ (1-x^2)\frac{d^2\phi_n(x)}{dx^2} - 3x\frac{d\phi_n(x)}{dx} \right\} = -t\sum_{n=0}^{\infty} \phi_n(x)\frac{d}{dt}\left\{ \frac{1}{t}\frac{d}{dt}t^{n+2} \right\}$$

$$= -\sum_{n=0}^{\infty} n(n+2)t^n \phi_n(x). \tag{6.7.11}$$

If we now equate the coefficients of powers of t on both sides of the equation we find

$$(1-x^2)\frac{d^2\phi_n(x)}{dx^2} - 3x\frac{d\phi_n(x)}{dx} + n(n+2)\phi_n(x) = 0. \tag{6.7.12}$$

In other words $\phi_n(x)$ satisfies Chebyshev's equation (6.2.1) and so must be multiples of the Chebyshev polynomials of the second kind. That is $\phi_n(x) = a_n U_n(x)$. We can work out a_n in each case by expanding $w(1,t)$ in powers of t for each of the functions on the right-hand side of Eqs. (6.7.2) and (6.7.4) - (6.7.6).

If we put $x = 1$ on the right-hand side of Eq. (6.7.2),

$$\sum_{n=0}^{\infty} t^n \phi_n(1) = \frac{1}{(1-t)^2} = 1 + 2t + 3t^2 + \dots$$

so that $\phi_n(1) = n + 1$ and that therefore $\phi_n(x) = U_n(x)$.

If we put $x = 1$ on the right-hand side of (6.7.4)

$$\sum_{n=0}^{\infty} t^n \phi_n(1) = \frac{1+t}{(1-t)^3} = 1 + 4t + 9t^2 + \dots + (n+1)^2 t^n + \dots$$

so in this case $\phi_n(1) = (n+1)^2$ and $\phi_n(x) = (n+1)U_n(x)$.

If we put $x = 1$ on the right-hand side of (6.7.5)

$$\sum_{n=0}^{\infty} t^n \phi_n(1) = \frac{1}{2(1-t)^{(3/2)}} = \frac{1}{2}\left(1 + \frac{3}{2}t + \frac{\frac{3}{2}\cdot\frac{5}{2}}{1.2}t^2 + \dots + \frac{\frac{3}{2}\dots(n+1/2)}{n!}t^n + \dots\right).$$

Therefore

$$\phi_n(1) = \frac{\Gamma(n+3/2)}{n!\,\Gamma(1/2)} \quad \text{so that} \quad \phi_n(x) = \frac{\Gamma(n+3/2)}{(n+1)!\,\Gamma(1/2)} U_n(x).$$

If we put $x = 1$ on the right-hand side of (6.7.6)

$$\sum_{n=0}^{\infty} t^n \phi_n(1) = -\frac{1}{2}\frac{1}{\sqrt{1-t}} = -\frac{1}{2}\left(1 + \frac{1}{2}t + \frac{\frac{1}{2}\cdot\frac{3}{2}}{2}t^2 + \frac{\frac{1}{2}\cdot\frac{3}{2}\dots(n-1/2)}{n!}t^n + \dots\right).$$

Therefore

$$\phi_n(1) = \frac{\Gamma(n+1/2)}{n!\,\Gamma(-1/2)} \quad \text{so that} \quad \phi_n(x) = \frac{\Gamma(n+1/2)}{(n+1)!\,\Gamma(-1/2)} U_n(x).$$

If we put $x = 1$ on the right-hand side of (6.7.7)

$$\sum_{n=0}^{\infty} t^{n+1} \phi_n(1) = \frac{t}{1-t} = t + t^2 + t^3 + \dots$$

Therefore $\phi_n(1) = 1$ and so $\phi_n(x) = U_n(x)/(n+1)$.

6.8 Recurrence Relations

There is a simple recurrence relation between the Chebyshev polynomials which is most easily derived using the trigonometric identity $\sin(A+B) = \sin A \cos B + \cos A \sin B$. Using this identity we can show that

$$U_{n+1}(x) + U_{n-1}(x) = 2xU_n(x). \tag{6.8.1}$$

This relation can also be derived by writing the generating function Eq. (6.7.2) in the form

$$(1 - 2xt + t^2) \sum_{n=0}^{\infty} t^n U_n(x) = 1$$

and equating the coefficients of t^n to zero.

More generally, noting that $\sin[(n + m + 1)\theta] + \sin[(n - m + 1)\theta] = 2\sin[(n + 1)\theta]\cos[m\theta]$, we see that

$$U_{n+m}(x) + U_{n-m}(x) = 2U_n(x)T_m(x), \tag{6.8.2}$$

where $T_m(x)$ is the Chebyshev function of the first kind and $T_m(\cos\theta) = \cos(m\theta)$.

6.9 Addition Formula

A number of relations involving Chebyshev polynomials of the first and second kinds are described in the chapter on Chebyshev polynomials of the first kind.

We can use the generating functions in sections 6.7 and 5.7

$$nU_n(x) = 2 \sum_{m=1}^{n} U_{n-m}(x)T_m(x). \tag{6.9.1}$$

Multiplying the generating functions Eq. (6.7.2) and Eq. (5.7.2) and using the generating function Eq. (6.7.5) produces

$$2 \sum_{p=0}^{\infty} t^p U_p(x) \sum_{q=0}^{\infty} t^q T_q(x) = 2\frac{1 - xt}{(1 - 2xt + t^2)^2} = \sum_{n=0}^{\infty} (2 + n)t^n U_n(x).$$

Equating the coefficients of t^n on both sides of this equation produces Eq. (6.9.1) after removing the term $U_n(x)T_0(x)$ from the sum.

A relation between the Chebyshev polynomials of the second, third and fourth kinds can also be proved using the generating functions of sections 6.7, 7.7 and 8.7.

$$(n + 1)U_n(x) = \sum_{m=0}^{n} V_m(x)W_{n-m}(x). \tag{6.9.2}$$

If we multiply the generating functions Eq. (7.7.2) and Eq. (8.7.2),

$$\sum_{p=0}^{\infty} t^p V_p(x) \sum_{q=0}^{\infty} t^q W_q(x) = \frac{1 - t^2}{(1 - 2xt + t^2)^2} = \sum_{n=0}^{\infty} (n + 1)t^n U_n(x)$$

from Eq. (6.7.4). Equating the coefficients of t^n on both sides gives Eq. (6.9.2).

Another relation between the Chebyshev polynomials of the second, third and fourth kinds is

$$U_{2n}(x) = V_n(x)W_n(x) = [U_n(x)]^2 - [U_{n-1}(x)]^2. \qquad (6.9.3)$$

This can easily be shown using the trigonometric representations:

$$\frac{\sin[(2n+1)\theta]}{\sin\theta} = \frac{2\sin[(n+1/2)\theta]\cos[(n+1/2)\theta]}{2\sin(\theta/2)\cos(\theta/2)}$$

$$= \frac{\big(\cos(2n\theta) - \cos[(2n+2)\theta]\big)}{2\sin^2\theta} = \frac{\sin^2[(n+1)\theta] - \sin^2[n\theta]}{\sin^2\theta}.$$

6.10   Differential Relations

We can deduce a relation for the derivative $U_n'(x)$ most easily by substituting $x = \cos\theta$.

$$(1 - x^2)U_n'(x) = -\frac{n}{2}U_{n+1}(x) + \frac{n+2}{2}U_{n-1}(x). \qquad (6.10.1)$$

There is a simple relationship between Chebyshev polynomials of the first and second kinds:

$$U_n(x) = T_{n+1}'(x)/(n+1). \qquad (6.10.2)$$

This is easily proved using the trigonometric representation whereby

$$T_n'(\cos\theta) = \frac{d}{d(\cos\theta)}\big(\cos(n\theta)\big) = n\frac{\sin(n\theta)}{\sin\theta} = nU_{n-1}(\cos\theta).$$

This relationship between the Chebyshev polynomials of the first and second kinds is a particular case of the relationship between Gegenbauer polynomials of order n and $n+1$.

Another differential relationship between Chebyshev polynomials of the first and second kinds is easily proved using the trigonometrical representations:

$$\frac{d}{dx}\left(\sqrt{1 - x^2}\,U_n(x)\right) = -(n+1)T_{n+1}(x)/\sqrt{1 - x^2}. \qquad (6.10.3)$$

6.11   Step Up and Step Down Operators

These operators can be easily derived from relations (6.8.1) and (6.9.1)

$$U_{n-1} = S_n^- U_n(x) = \frac{1}{n+1}\left\{ nx + (1 - x^2)\frac{d}{dx} \right\} U_n(x) \qquad (6.11.1)$$

and

$$U_{n+1}(x) = S_n^+ U_n(x) = \frac{1}{n+1}\left\{ (n+2)x - (1 - x^2)\frac{d}{dx} \right\} U_n(x). \qquad (6.11.2)$$

References

Doman B G S, International Journal of Pure and Applied Mathematics, 2010, 63, pp. 197-205.

Ferrar W L, Textbook of Convergence, Oxford University Press, 1938.

Hochstrasser Urs W, Orthogonal Polynomials, Chapter 22, Handbook of Mathematical Functions, Eds. Abramowitz M and Stegun I A, Dover, 1970.

Koornwinder T H, Wong R, Koekoek R and Swarttouw R F, Chapter 18, NIST Handbook of Mathematical Functions, Eds. Olver W J, Lozier D W, Boisvert R F and Clark C W, NIST and Cambridge University Press, 2009.

Mason J C and Handscomb D C, Chebyshev Polynomials, Chapman & Hall, 2002.

Phillips E G, A Course of Analysis, Cambridge University Press, 1930.

Chapter 7

Chebyshev Polynomials of the Third Kind

7.1 Introduction

Chebyshev polynomials of the third kind $V_n(x)$ can be obtained using the Gram-Schmidt orthogonalisation process for polynomials in the domain $(-1, 1)$ with the weight factor $\sqrt{(1+x)/(1-x)}$. The resulting polynomial $R_n(x)$ is multiplied by a number which makes the value at $x = 1$ equal to 1. The resulting polynomials, $V_n(x)$ are multiples of the Jacobi polynomials $P_n^{(-1/2,1/2)}(x)$. In fact

$$V_n(x) = \frac{n!\sqrt{\pi}}{\Gamma(n+1/2)} P_n^{(-1/2,1/2)}(x). \qquad (7.1.1)$$

There is a simple relation between the Chebyshev polynomials of the third and second kinds:

$$V_n(x) = U_n(x) - U_{n-1}(x). \qquad (7.1.2)$$

This is most easily proved using the trigonometrical representation in section 7.4.

The first few Chebyshev polynomials of the third kind are:

$V_0(x) = 1.$

$V_1(x) = 2x - 1.$

$V_2(x) = 4x^2 - 2x - 1.$

$V_3(x) = 8x^3 - 4x^2 - 4x + 1.$

$V_4(x) = 16x^4 - 8x^3 - 12x^2 + 4x + 1.$

$V_5(x) = 32x^5 - 16x^4 - 32x^3 + 12x^2 + 6x - 1.$

$V_6(x) = 64x^6 - 32x^5 - 80x^4 + 32x^3 + 24x^2 - 6x - 1.$

$V_7(x) = 128x^7 - 64x^6 - 192x^5 + 80x^4 + 80x^3 - 24x^2 - 8x + 1.$

$V_8(x) = 256x^8 - 128x^7 - 448x^6 + 192x^5 + 240x^4 - 80x^3 - 40x^2 + 8x + 1.$

Graphs of $V_n(x)$ for $n = 1$ to $n = 4$.

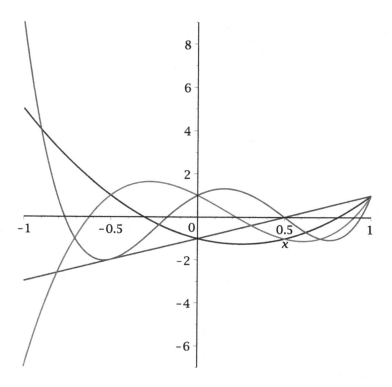

7.2 Differential Equation

The ordinary differential equation satisfied by Chebyshev polynomials of the third kind is

$$(1 - x^2)\frac{d^2y}{dx^2} + (1 - 2x)\frac{dy}{dx} + \lambda y = 0. \tag{7.2.1}$$

The point $x = 0$ is an ordinary point. This means that we can express the solution in the form of a power series $y = \sum_{n=0}^{\infty} a_n x^n$. The singularities of the differential equation are at $x = \pm 1$ and so the radius of convergence of the power series will be 1. On substituting this power series into the

diferential equation we find

$$(1 - x^2) \sum_{n=0}^{\infty} n(n-1)a_n x^{n-2} + (1 - 2x) \sum_{n=0}^{\infty} n a_n x^{n-1} + \lambda \sum_{n=0}^{\infty} a_n x^n = 0.$$

If we equate the coefficients of x^n to zero we get

$$(n+2)(n+1)a_{n+2} + (n+1)a_{n+1} + (\lambda - n(n+1))a_n = 0.$$

This leads to the recurrence relation

$$a_{n+2} = -\frac{1}{n+2} a_{n+1} + \frac{n(n+1) - \lambda}{(n+1)(n+2)} a_n. \tag{7.2.2}$$

This gives us two linearly independent series solutions each containing both odd and even powers of x.

One of the series solutions will terminate and become an mth order polynomial if $a_m = -2a_{m-1}$ and $\lambda = m(m+1)$ where m is an integer. This condition determines all of the coefficients of this polynomial solution up to a multiplicative constant.

We are now going to show that the polynomial solutions $R_m(x)$ satisfy the orthogonality relations for the Chebyshev polynomials of the third kind and must therefore be multiples of them.

7.3 Orthogonality

The differential equation (7.2.1) for $R_m(x)$ can be written in Sturm Liouville form

$$\frac{d}{dx}\left((1+x)^{3/2}(1-x)^{1/2} \frac{dR_m(x)}{dx} \right) = -m(m+1)\sqrt{\frac{1+x}{1-x}} R_m(x). \tag{7.3.1}$$

If we multiply this by $R_n(x)$, where $n \neq m$ and integrate from -1 to 1, the left-hand side becomes

$$\int_{-1}^{1} R_n(x) \frac{d}{dx}\left((1-x)^{3/2}(1+x)^{1/2} \frac{dR_m(x)}{dx} \right) dx.$$

On integrating by parts and noting that the integrated term vanishes at both end points $x = \pm 1$, we find

$$\int_{-1}^{1} (1+x)^{3/2}(1-x)^{1/2} R_n'(x) R_m'(x) dx$$

$$= m(m+1) \int_{-1}^{1} \sqrt{\frac{1+x}{1-x}} R_n(x) R_m(x) dx.$$

If we follow the same procedure as above but with m and n interchanged, we will produce an equation which is the same as that above on the left-hand side and the same integral on the right but with a coefficient $n(n+1)$ instead of $m(m+1)$. If we subtract one of these equations from the other, we get

$$[m(m+1) - n(n+1)] \int_{-1}^{1} \sqrt{\frac{1+x}{1-x}} V_n(x) V_m(x) dx = 0,$$

and so for $n \neq m$

$$\int_{-1}^{1} \sqrt{\frac{1+x}{1-x}} V_n(x) V_m(x) dx = 0. \tag{7.3.2}$$

The polynomials $R_m(x)$ thus satisfy the same orthogonality relations as the Chebyshev polynomials of the third kind and so will be multiples of them. If we divide $R_m(x)$ by $R_m(1)$ so that the resulting polynomial equals 1 when $x = 1$ and so will be the Chebyshev polynomial of the third kind.

Since x^n can be represented as a linear combination of the Chebyshev polynomials of the third kind $V_p(x)$ for $0 \leq p \leq n$, we can deduce that

$$\int_{-1}^{1} \sqrt{\frac{1+x}{1-x}} x^n V_m(x) dx = 0 \qquad \text{for all } n < m. \tag{7.3.3}$$

7.4 Trigonometric Representaion

If we change the independent variable in the differential equation (7.2.1) to θ by writing $x = \cos\theta$ and $\lambda = n(n+1)$, the equation for $g(\theta) = \cos(\theta/2) y(\cos\theta)$ becomes

$$\frac{d^2 g}{d\theta^2} + \left(n + \frac{1}{2}\right)^2 g = 0. \tag{7.4.1}$$

The solutions of this equation are $\cos[(n+1/2)\theta]$ and $\sin[(n+1/2)\theta]$. The solution $y = \sin[(n+1/2)\theta]/\cos(\theta/2)$ is singular at $\theta = \pi$ and therefore cannot be represented by a polynomial in $\cos\theta$. The acceptable polynomial solution is then $R_n(\cos\theta) = \cos[(n+1/2)\theta]/\cos(\theta/2)$. If we now let $\theta \to 0$ which corresponds to $x \to 1$ we see that $R_n(1) = 1$ and so

$$V_n(\cos\theta) = \frac{\cos[(n+1/2)\theta]}{\cos(\theta/2)}. \tag{7.4.2}$$

Equation (7.1.2) follows from the observation that

$$\frac{\cos[(n+1/2)\theta]}{\cos(\theta/2)} = \frac{\sin[(n+1)\theta] - \sin(n\theta)}{2\sin(\theta/2)\cos(\theta/2)}.$$

The orthogonality relation can be proved directly by substituting $x = \cos\theta$ and using the representation $V_n(\cos\theta) = \cos[(n+1/2)\theta]/\cos(\theta/2)$ in Eq. (7.3.2) which gives

$$\int_0^\pi \cos[(n+1/2)\theta]\cos[(m+1/2)\theta]d\theta = 0 \qquad n \neq m. \qquad (7.4.3)$$

$$h_n = \int_{-1}^1 \sqrt{\frac{1+x}{1-x}}[V_n(x)]^2 dx = 2\int_0^\pi [\cos(n+1/2)\theta]^2 d\theta = \pi. \qquad (7.4.4)$$

The coefficient of x^n in $V_n(x)$ can be obtained using Euler's formula for $\cos\theta$. Writing $z = e^{i\theta/2}$

$$z^{2n+1} + z^{-(2n+1)} = \left(z + \frac{1}{z}\right)\left(z^{2n} - z^{2n-2} + \dots z^{-(2n-2)} + z^{-2n}\right).$$

Therefore

$$\cos[(n+1/2)\theta] = 2\cos(\theta/2)[\cos n\theta - \cos(n-1)\theta + \dots]$$

and noting that $2^{n-1}[\cos\theta]^n = \cos n\theta + n\cos(n-2)\theta + \dots$. Thus

$$\frac{\cos[(n+1/2)\theta]}{\cos\theta/2} = 2^n[\cos\theta]^n + O(\cos\theta)^{n-1}.$$

Therefore the coefficient of x^n in $V_n(x)$, k_n is given by

$$k_n = 2^n. \qquad (7.4.5)$$

We can obtain the cofficient of x^{n-1}, k_n', by noting that from section (7.2), $k_n = -2k_{n-1}'$. Hence

$$k_n' = -2^{n-1}. \qquad (7.4.6)$$

7.5 Rodrigues Formula

By treating $(1+x)^{m+1/2}(1-x)^{m-1/2}$ as the product of its two factors, we see that if $n < m$ the nth derivative of $(1+x)^{m+1/2}(1-x)^{m-1/2}$ contains the factor $(1+x)^{m-n+1/2}(1-x)^{m-n-1/2}$ and therefore that

$$\frac{d^n}{dx^n}\left\{\sqrt{\frac{1+x}{1-x}}(1-x^2)^m\right\}$$

$$= \frac{d^n}{dx^n}\left\{(1-x)^{m-1/2}(1+x)^{m+1/2}\right\} = 0 \quad \text{when} \quad x = \pm 1.$$

From this result, it follows that

$$\int_{-1}^{1} \frac{d^m}{dx^m}\left\{\sqrt{\frac{1+x}{1-x}}(1-x^2)^m\right\}dx = 0.$$

If we use the method of integration by parts we see that

$$\int_{-1}^{1} x\frac{d^m}{dx^m}\left\{\sqrt{\frac{1+x}{1-x}}(1-x^2)^m\right\}dx = 0 \qquad m > 1$$

and using the method of integration by parts n times we see that

$$\int_{-1}^{1} x^n \frac{d^m}{dx^m}\left\{\sqrt{\frac{1+x}{1-x}}(1-x^2)^m\right\}dx = 0 \quad \text{provided that} \quad m > n. \quad (7.5.1)$$

Let us define the m th order polynomial $Q_m(x)$ by

$$Q_m(x) = \sqrt{\frac{1-x}{1+x}}\frac{d^m}{dx^m}\left\{\sqrt{\frac{1+x}{1-x}}(1-x^2)^m\right\}. \qquad (7.5.2)$$

Then

$$\int_{-1}^{1} \sqrt{\frac{1+x}{1-x}}x^n Q_m(x)dx = 0 \qquad n < m.$$

In other words, the m th order polynomial $Q_m(x)$ is orthogonal to x^n for all values of $n < m$. This means that $Q_m(x)$ is orthogonal to $Q_n(x)$ for all values of $n < m$. These polynomials must therefore be multiples of the Chebyshev polynomials we found earlier. We can find out what this multiple is by evaluating $Q_n(1)$ by writing

$$Q_n(1) = \lim_{x \to 1} \sqrt{\frac{1-x}{1+x}}\frac{d^n}{dx^n}\left\{(1-x)^{n-1/2}(1+x)^{n+1/2}\right\}.$$

The only term which contributes in the limit $x \to 1$ is $(1+x)^{n+1/2}$ multiplied by the n th derivative of $(1-x)^{n-1/2}$. Thus

$$Q_n(1) = (n-1/2)(n-3/2)...(1/2)2^n(-1)^n. \qquad (7.5.3)$$

Thus

$$V_n(x) = \frac{(-1)^n}{1 \cdot 3 \cdot 5...(2n-1)}\sqrt{\frac{1-x}{1+x}}\frac{d^n}{dx^n}\left\{(1-x)^{n-1/2}(1+x)^{n+1/2}\right\}. \quad (7.5.4)$$

7.6 Explicit Expression

The simplest way to obtain a power series expansion for $V_n(x)$ is to use Eq. (7.1.2)

$$V_n(x) = U_n(x) - U_{n-1}(x)$$

$$= \sum_{q=0}^{[n/2]} (-1)^q \frac{(n-q)!}{q!\,(n-2q)!} \, (2x)^{n-2q} - \sum_{q=0}^{[(n-1)/2]} (-1)^q \frac{(n-q-1)!}{q!\,(n-1-2q)!} \, (2x)^{n-1-2q},$$

(7.6.1)

where $[n/2]$ is the largest integer less than or equal to $n/2$.

We can express x^n in terms of the Chebyshev polynomials of the third kind

$$x^n = \sum_{m=0}^{n} a_m V_m(x). \tag{7.6.2}$$

The simplest way to evaluate the coefficients a_m is to use the trigonometric representation. Noting that

$$(e^{i\theta} + e^{-i\theta})^n (e^{i\theta/2} + e^{-i\theta/2}) = \sum_{m=0}^{n} \binom{n}{m} \left(e^{(n+1/2-2m)i\theta} + e^{(n-1/2-2m)i\theta} \right).$$

If we replace m by $n - m$ in the second exponent on the right-hand side, we get

$$2^n \cos^n \theta \cos \theta/2 = \sum_{m=0}^{n} \binom{n}{m} \cos[(n - 2m + 1/2)\theta]$$

and so

$$x^n = 2^{-n} \sum_{m=0}^{[n/2]} \binom{n}{m} V_{n-2m} + 2^{-n} \sum_{m=0}^{[(n-1)/2]} \binom{n}{m} V_{n-1-2m} \tag{7.6.3}$$

where $[n/2]$ is the largest integer less than or equal to $n/2$.

An alternative way to evaluate the expansion coefficients a_m is to multiply equation Eq. (7.6.4) by $V_q(x)$ and integrate from -1 to 1. Making use of the orthogonality property of the Chebyshev polynomials, we see

$$a_q h_q = \int_{-1}^{1} x^n V_q(x) \sqrt{\frac{1+x}{1-x}} \, dx = \frac{(-2)^q \, q!}{(2q)!} \int_{-1}^{1} x^n \frac{d^q}{dx^q} \left[(x^2-1)^q \sqrt{\frac{1+x}{1-x}} \right] dx$$

$$= \frac{2^q \, q! \, n!}{(2q)! \, (n-q)!} \int_{-1}^{1} x^{n-q} (1+x)(1-x^2)^{q-1/2} dx$$

after integrating by parts q times with the integrated parts being zero at the end points. Looking at the factor $(1 + x)$, we see that if $n - q$ is even, the contribution of the x part is zero since the integrand with x is odd. Similarly, the contribution of 1 will be zero if $n - q$ is odd. We therefore split the integral into two parts writing $q = n - 2m$ in one part and $q = n - 1 - 2m$ in the other. Then, with $h_q = \pi$

$$\pi a_{n-2m} = \frac{2^{n-2m} \, n! \, (n - 2m)!}{(2n - 4m)! \, (2m)!} \int_{-1}^{1} x^{2m} (1 - x^2)^{n-2m-1/2} dx.$$

$$\pi a_{n-1-2m} = \frac{2^{n-1-2m} \, n! \, (n - 1 - 2m)!}{(2n - 4m - 2)! \, (2m + 1)!} \int_{-1}^{1} x^{2m+2} (1 - x^2)^{n-2m-3/2} dx.$$

Writing $t = x^2$ converts the first integral into the Beta function

$$B(m + 1/2, n - 2m + 1/2) = \frac{\Gamma(m + 1/2)\Gamma(n - 2m + 1/2)}{\Gamma(n - m + 1)}$$

and the second integral to

$$B(m + 3/2, n - 2m - 1/2) = \frac{\Gamma(m + 3/2)\Gamma(n - 2m - 1/2)}{\Gamma(n - m + 1)}.$$

(See the General Appendix for the properties of the Beta function.) Then

$$x^n = \sum_{m=0}^{[n/2]} \frac{2^{-n} \, n!}{m! \, (n - m)!} V_{n-2m}(x) + \sum_{m=0}^{[(n-1)/2]} \frac{2^{-n} \, n!}{m! \, (n - m)!} V_{n-1-2m}(x).$$

7.7 Generating Functions

The simplest generating function can be derived by taking the real part of the identity

$$\sum_{n=0}^{\infty} t^n e^{(n+1/2)i\theta} = \frac{e^{i\theta/2}}{1 - te^{i\theta}} \qquad |t| < 1 \tag{7.7.1}$$

and putting $x = \cos\theta$.

This leads to the relation

$$\sum_{n=0}^{\infty} t^n \cos[(n + 1/2)\theta] = \cos(\theta/2) \frac{1 - t}{1 - 2t\cos\theta + t^2}.$$

Dividing through by $\cos(\theta/2)$ and putting $x = \cos\theta$, we have from the trigonometric representation

$$\sum_{n=0}^{\infty} t^n V_n(x) = \frac{1 - t}{1 - 2xt + t^2}. \tag{7.7.2}$$

The derivation of this generating function is a special case of using the identity

$$\sum_{n=0}^{\infty} \frac{\Gamma(n+\mu)}{n!\,\Gamma(\mu)} t^n e^{i(n+1/2)\theta} = \frac{e^{i\theta/2}}{(1-te^{i\theta})^\mu}, \tag{7.7.3}$$

and taking the real part.

We can derive other generating functions by choosing different values for μ. If we let $\mu = 2$,

$$\sum_{n=0}^{\infty}(n+1)t^n \cos[(n+1/2)\theta] = Re\left(\frac{e^{i\theta/2}}{(1-te^{i\theta})^2}\right)$$

$$= \frac{\cos(\theta/2)[1-2t+t^2(2x-1)]}{R^4}$$

where $R = \sqrt{1-2xt+t^2}$. Therefore

$$\sum_{n=0}^{\infty}(n+1)t^n V_n(x) = \frac{1-2t+t^2(2x-1)}{R^4}. \tag{7.7.4}$$

We can derive another generating function if we let $\mu = 1/2$

$$\sum_{n=0}^{\infty} \frac{\Gamma(n+1/2)}{n!\,\Gamma(1/2)} t^n \cos[(n+1/2)\theta] = Re\left(\frac{e^{i\theta/2}}{\sqrt{1-te^{i\theta}}}\right)$$

$$= Re\left[\frac{e^{i\theta/2}}{\sqrt{2R}}\left(\sqrt{1-xt+R} + \frac{it\sin\theta}{\sqrt{1-xt+R}}\right)\right] = \frac{\cos(\theta/2)(1-t+R)}{R\sqrt{2(1-xt+R)}}.$$

Therefore

$$\sum_{n=0}^{\infty} \frac{\Gamma(n+1/2)}{n!\,\Gamma(1/2)} t^n V_n(x) = \frac{1-t+R}{R\sqrt{2(1-xt+R)}}. \tag{7.7.5}$$

For $\mu = -1/2$

$$\sum_{n=0}^{\infty} \frac{\Gamma(n-1/2)}{n!\,\Gamma(-1/2)} t^n \cos[(n+1/2)\theta] = Re\left(e^{i\theta/2}\sqrt{1-te^{i\theta}}\right)$$

$$= Re\left[\frac{e^{i\theta/2}}{\sqrt{2}}\left(\sqrt{1-xt+R} - \frac{it\sin\theta}{\sqrt{1-xt+R}}\right)\right]$$

$$= \frac{\cos(\theta/2)(1-2xt+t+R)}{\sqrt{2(1-xt+R)}}.$$

Therefore

$$\sum_{n=0}^{\infty} \frac{\Gamma(n-1/2)}{n!\,\Gamma(-1/2)} t^n V_n(x) = \frac{1 - 2xt + t + R}{\sqrt{2(1 - xt + R)}}. \qquad (7.7.6)$$

An alternative derivation of the validity of these generating functions is to define functions $\phi_n(x)$ by

$$\sum_{n=0}^{\infty} t^n \phi_n(x) = w(x,t), \qquad (7.7.7)$$

for each function $w(x,t)$ on the right-hand side of (7.7.2), (7.7.4), (7.7.5) or (7.7.6). We can see in each case that $\phi_n(x)$ is an m th order polynomial in x by using the binomial theorem to expand the appropriate $w(x,t)$ in powers of t and collecting the terms in t^n. The next step is to show that for each possible function $w(x,t)$,

$$(1 - x^2)\frac{\partial^2}{\partial x^2}w(x,t) + (1 - 2x)\frac{\partial}{\partial x}w(x,t) = -t\frac{\partial^2}{\partial t^2}\Big(tw(x,t)\Big). \qquad (7.7.8)$$

This can be most easily done using a computer algebra package such as Maple.

If we substitute the left-hand side of (7.7.7) into (7.7.8) then

$$\sum_{n=0}^{\infty} t^n \left\{ (1 - x^2)\frac{d^2\phi_n(x)}{dx^2} + (1 - 2x)\frac{d\phi_n(x)}{dx} \right\}$$

$$= -t\sum_{n=0}^{\infty} \phi_n(x)\frac{d^2}{dt^2}t^{n+1} = -\sum_{n=0}^{\infty} n(n+1)t^n\phi_n(x). \qquad (7.7.9)$$

On equating powers of t on both sides of this equation we see that $\phi_n(x)$ satisfies the Chebyshev differential equation and must therefore be a multiple of the Chebyshev polynomial of the third kind, that is $a_n V_n(x)$. We can find the value of a_n for each generating function by putting $x = 1$ in each of the expressions on the right-hand side of Eqs. (7.7.2), (7.7.4), (7.7.5) and (7.7.6) and expanding in powers of t.

For generating function (7.7.2),

$$w(1,t) = \frac{1}{1-t} = 1 + t + t^2 + t^3 + \dots$$

so $a_n = 1$.

For generating function (7.7.4),

$$w(1,t) = \frac{1}{(1-t)^2} = 1 + 2t + 3t^2 + \dots$$

and so $a_n = n + 1$.

For generating function (7.7.5),

$$w(1,t) = \frac{1}{\sqrt{1-t}} = 1 + \frac{1}{2}t + \frac{\frac{1}{2} \cdot \frac{3}{2}}{1.2}t^2 + ...,$$

and so $a_n = \Gamma(n + 1/2)/[n!\Gamma(1/2)] = (2n)!/[2^{2n}(n!)^2]$.

For generating function (7.7.6),

$$w(1,t) = \sqrt{1-t} = 1 - \frac{1}{2}t - \frac{\frac{1}{2} \cdot \frac{1}{2}}{1.2}t^2 - ...,$$

and so $a_n = \Gamma(n - 1/2)/[n!\,\Gamma(-1/2)] = -(2n - 2)!/[2^{2n-1}n!\,(n - 1)!]$.

7.8 Recurrence Relations

There is a simple recurrence relation between the Chebyshev polynomials which is most easily derived using the trigonometric identity $\cos(A + B) = \cos A \cos B - \sin A \sin B$.

$$V_{n+1}(x) + V_{n-1}(x) = 2xV_n(x). \tag{7.8.1}$$

This relation can also be simply derived by writing the generating function Eq. (7.7.2) in the form

$$(1 - 2xt + t^2)\sum_{n=0}^{\infty} t^n V_n(x) = 1 - t.$$

A more general recurrence relation which can be derived using the trigonometric representation is

$$V_{n+m}(x) + V_{n-m}(x) = 2T_m(x)V_n(x), \tag{7.8.2}$$

where $T_m(x)$ is a Chebyshev polynomial of the first kind.

7.9 Differential Relation

We can derive a relation for the derivative $V_n'(x)$ by substituting $x = \cos\theta$.

$$(1 - x^2)V_n'(x) = \frac{1}{2}\left(n + \frac{1}{2}\right)\left(V_{n-1}(x) - V_{n+1}(x)\right) - \frac{1}{2}(1 - x)V_n(x). \tag{7.9.1}$$

7.10 Step Up and Step Down Operators

These operators can be easily derived from relations (7.8.1) and (7.9.1).

$$V_{n-1}(x) = S_n^- V_n(x) = \frac{2}{2n+1}\left\{nx + 1/2 + (1-x^2)\frac{d}{dx}\right\}V_n(x) \quad (7.10.1)$$

and

$$V_{n+1}(x) = S_n^+ V_n(x) = \frac{2}{2n+1}\left\{(n+1)x - 1/2 - (1-x^2)\frac{d}{dx}\right\}V_n(x). \quad (7.10.2)$$

References

Doman B G S, International Journal of Pure and Applied Mathematics, 2010, 63, pp. 197-205.

Ferrar W L, Textbook of Convergence, Oxford University Press, 1938.

Hochstrasser Urs W, Orthogonal Polynomials, Chapter 22, Handbook of Mathematical Functions, Eds. Abramowitz M and Stegun I A, Dover, 1970.

Koornwinder T H, Wong R, Koekoek R and Swarttouw R F, Chapter 18, NIST Handbook of Mathematical Functions, Eds. Olver W J, Lozier D W, Boisvert R F and Clark C W, NIST and Cambridge University Press, 2009.

Mason J C and Handscomb D C, Chebyshev Polynomials, Chapman & Hall, 2002.

Phillips E G, A Course of Analysis, Cambridge University Press, 1930.

Chapter 8

Chebyshev Polynomials of the Fourth Kind

8.1 Introduction

Chebyshev polynomials of the fourth kind $W_n(x)$ can be obtained using the Gram-Schmidt orthogonalisation process for polynomials in the domain $(-1, 1)$ with the weight factor $\sqrt{(1-x)/(1+x)}$. The resulting polynomial $R_n(x)$ is multiplied by a number which makes the value at $x = 1$ equal to $2n + 1$. The resulting polynomials, $W_n(x)$ are multiples of the Jacobi polynomials $P_n^{(1/2, -1/2)}(x)$. In fact

$$W_n(x) = \frac{n!\sqrt{\pi}}{\Gamma(n + 1/2)} P_n^{(1/2, -1/2)}(x). \tag{8.1.1}$$

There is a simple relation between the Chebyshev polynomials of the fourth and second kinds:

$$W_n(x) = U_n(x) + U_{n-1}(x). \tag{8.1.2}$$

This is most easily proved using the trigonometrical representation in section 8.4.

The first few Chebyshev polynomials of the fourth kind are:

$W_0(x) = 1.$
$W_1(x) = 2x + 1.$
$W_2(x) = 4x^2 + 2x - 1.$
$W_3(x) = 8x^3 + 4x^2 - 4x - 1.$
$W_4(x) = 16x^4 + 8x^3 - 12x^2 - 4x + 1.$
$W_5(x) = 32x^5 + 16x^4 - 32x^3 - 12x^2 + 6x + 1.$
$W_6(x) = 64x^6 + 32x^5 - 80x^4 - 32x^3 + 24x^2 + 6x - 1.$
$W_7(x) = 128x^7 + 64x^6 - 192x^5 - 80x^4 + 80x^3 + 24x^2 - 8x - 1.$
$V_8(x) = 256x^8 + 128x^7 - 448x^6 - 192x^5 + 240x^4 + 80x^3 - 40x^2 - 8x + 1.$

Graphs of $W_n(x)$ for $n = 1$ to $n = 4$.

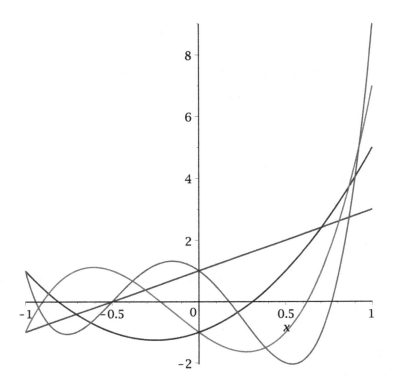

8.2 Differential Equation

The ordinary differential equation satisfied by Chebyshev polynomials of the fourth kind is

$$(1 - x^2)\frac{d^2y}{dx^2} - (1 + 2x)\frac{dy}{dx} + \lambda y = 0. \tag{8.2.1}$$

The point $x = 0$ is an ordinary point. This means that we can express the solution in the form of a power series in $y = \sum a_n x^n$. The singularities of the differential equation are at $x = \pm 1$ and so the radius of convergence of the power series will be 1. On substituting this power series into the differential equation we find

$$(1 - x^2) \sum_{n=0}^{\infty} n(n-1)a_n x^{n-2} - (1 + 2x) \sum_{n=0}^{\infty} n a_n x^{n-1} + \lambda \sum_{n=0}^{\infty} a_n x^n = 0.$$

If we equate the coefficients of x^n to zero we get

$$(n + 2)(n + 1)a_{n+2} - (n + 1)a_{n+1} + (\lambda - n(n + 1))a_n = 0.$$

This leads to the recurrence relation

$$a_{n+2} = \frac{1}{n+2} a_{n+1} + \frac{n(n+1) - \lambda}{(n+2)(n+1)} a_n. \tag{8.2.2}$$

This gives us two linearly independent series solutions, each containing both even and odd powers of x.

One of these series will terminate and become an m th order polynomial if $a_m = 2a_{m-1}$ and $\lambda = m(m + 1)$ where m is an integer. This condition determines all of the coefficients of this polynomial solution up to a multiplicative constant.

We are now going to show that the polynomial solutions $R_m(x)$ satisfy the orthogonality relations for the Chebyshev polynomials of the fourth kind and must therefore be multiples of them.

8.3 Orthogonality

The differential equation (8.2.1) for $R_m(x)$ can be written in Sturm Liouville form

$$\frac{d}{dx} \left((1 - x)^{3/2}(1 + x)^{1/2} \frac{dR_m(x)}{dx} \right) = -m(m + 1)\sqrt{\frac{1 - x}{1 + x}} R_m(x). \tag{8.3.1}$$

If we multiply this by $R_n(x)$, where $n \neq m$ and integrate from -1 to 1 the left-hand side becomes

$$\int_{-1}^{1} R_n(x) \frac{d}{dx} \left((1 - x)^{3/2}(1 + x)^{1/2} \frac{dR_m(x)}{dx} \right).$$

On integrating by parts and noting that the integrated term vanishes at both end points $x = \pm 1$, we find that

$$\int_{-1}^{1} (1 - x)^{3/2}(1 + x)^{1/2} R_n'(x) R_m'(x) dx$$

$$= m(m + 1) \int_{-1}^{1} \sqrt{\frac{1 - x}{1 + x}} R_n(x) R_m(x) dx.$$

If we follow the same procedure as above but with m and n interchanged, we will produce an equation which is the same as that above on the left-hand side and with the same integral on the right but with a coefficient $n(n + 1)$ instead of $m(m + 1)$. If we subtract one of these equations from the other, we get

$$[m(m + 1) - n(n + 1)] \int_{-1}^{1} \sqrt{\frac{1 - x}{1 + x}} R_n(x) R_m(x) dx = 0,$$

and so for $n \neq m$

$$\int_{-1}^{1} \sqrt{\frac{1 - x}{1 + x}} R_n(x) R_m(x) dx = 0. \tag{8.3.2}$$

The polynomials $R_m(x)$ thus satisfy the same orthogonality relations as the Chebyshev polynomials of the fourth kind and so will be multiples of them. If we multiply $R_m(x)$ by $(2m + 1)/R_m(1)$ the resulting polynomial will equal $2m + 1$ when $x = 1$ and so will be the Chebyshev polynomial of the fourth kind.

Since x^n can be represented as a linear combination of the Chebyshev polynomials of the fourth kind $W_p(x)$ for $0 \leq p \leq n$, we can deduce that

$$\int_{-1}^{1} \sqrt{\frac{1 - x}{1 + x}} x^n W_m(x) dx = 0 \qquad \text{for all } n < m. \tag{8.3.3}$$

8.4   Trigonometric Representation

If we change the independent variable in the differential equation (8.2.1) to θ by putting $x = \cos\theta$ and $\lambda = n(n + 1)$, the equation for $g(\theta) = \sin(\theta/2)y(\cos\theta)$ becomes

$$\frac{d^2 g}{d\theta^2} + \left(n + \frac{1}{2}\right)^2 g = 0. \tag{8.4.1}$$

The solutions of this equation are $\cos[(n + 1/2)\theta]$ and $\sin[(n + 1/2)\theta]$. The solution $y = \cos[(n + 1/2)\theta]/\sin(\theta/2)$ is singular at $\theta = 0$ and therefore cannot be represented by a polynomial in $\cos\theta$. The acceptable polynomial solution is then $R_n(\cos\theta) = \sin[(n + 1/2)\theta]/\sin(\theta/2)$. If we now let $\theta \to 0$ which corresponds to $x \to 1$ we see that $R_n(1) = 2n + 1$ and so

$$W_n(\cos\theta) = \frac{\sin[(n + 1/2)\theta]}{\sin(\theta/2)}. \tag{8.4.2}$$

Equation (8.1.2) follows from the observation that

$$\frac{\sin[(n + 1/2)\theta]}{\sin(\theta/2)} = \frac{\sin[(n + 1)\theta] - \sin(n\theta)}{2\sin(\theta/2)\cos(\theta/2)}.$$

The orthogonality relation can be proved by substituting $x = \cos\theta$ and using the trigonometric representation $W_n(\cos\theta) = \sin[(n+1/2)\theta]/\sin(\theta/2)$ in Eq. (2.2) which gives

$$\int_0^\pi \sin[(n+1/2)\theta]\sin[(m+1/2)\theta]d\theta = 0 \qquad n \neq m. \tag{8.4.3}$$

$$h_n = \int_{-1}^1 \sqrt{\frac{1-x}{1+x}}\,[W_n(x)]^2dx = 2\int_0^\pi [\sin(n+1/2)\theta]^2d\theta = \pi. \tag{8.4.4}$$

The coefficient of x^n in $W_n(x)$ can be obtained using Euler's formula for $\sin\theta$. Writing $z = e^{i\theta/2}$,

$$z^{2n+1} - z^{-(2n+1)} = \left(z - \frac{1}{z}\right)\left(z^{2n} + z^{2n-2} + ... + z^{-(2n-2)} + z^{-2n}\right).$$

Therefore

$$\sin(n+1/2)\theta = 2\sin(\theta/2)(\cos n\theta + \cos[(n-1)\theta] + ...)$$

and noting that $2^{n-1}[\cos\theta]^n = \cos n\theta + n\cos[(n-2)\theta] + ...$. Thus

$$\frac{\sin(n+1/2)\theta}{\sin\theta/2} = 2^n[\cos\theta]^n + O(\cos\theta)^{n-1}.$$

Therefore the coefficient of x^n in $W_n(x)$,

$$k_n = 2^n. \tag{8.4.5}$$

We can obtain the coefficient of x^{n-1}, k'_n, by noting from sectio (8.2), that $k_n = 2k'_{n-1}$. Hence

$$k'_n = 2^{n-1}. \tag{8.4.6}$$

8.5 Rodrigues Formula

By treating $(1+x)^{m-1/2}(1-x)^{m+1/2}$ as the product of its two factors, we see that if $n < m$, the nth derivative of $(1+x)^{m-1/2}(1-x)^{m+1/2}$ contains the factor $(1+x)^{m-n-1/2}(1-x)^{m-n+1/2}$ and therefore that

$$\frac{d^n}{dx^n}\left\{\sqrt{\frac{1-x}{1+x}}(1-x^2)^m\right\}$$

$$= \frac{d^n}{dx^n}\left\{(1-x)^{m+1/2}(1+x)^{m-1/2}\right\} = 0 \quad \text{when} \quad x = \pm 1.$$

From this result, it follows that

$$\int_{-1}^{1} \frac{d^m}{dx^m}\left\{ \sqrt{\frac{1-x}{1+x}}(1-x^2)^m \right\} dx = 0.$$

If we integrate by parts we see that

$$\int_{-1}^{1} x \frac{d^m}{dx^m}\left\{ \sqrt{\frac{1-x}{1+x}}(1-x^2)^m \right\} dx = 0 \qquad m > 1$$

and integrating by parts n times we see that

$$\int_{-1}^{1} x^n \frac{d^m}{dx^m}\left\{ \sqrt{\frac{1-x}{1+x}}(1-x^2)^m \right\} dx = 0 \qquad \text{provided that } m > n. \quad (8.5.1)$$

Let us define the mth order polynomial $Q_m(x)$ by

$$Q_m(x) = \sqrt{\frac{1+x}{1-x}} \frac{d^m}{dx^m}\left\{ \sqrt{\frac{1-x}{1+x}}(1-x^2)^m \right\}. \qquad (8.5.2)$$

Then

$$\int_{-1}^{1} \sqrt{\frac{1-x}{1+x}} x^n Q_m(x) dx = 0 \qquad n < m.$$

In other words, the mth order polynomial $Q_m(x)$ is orthogonal to x^n for all values of $n < m$. This means that $Q_m(x)$ is orthogonal to $Q_n(x)$ for all values of $n < m$. These polynomials must therefore be multiples of the Chebyshev polynomials we found earlier. We can find out what this multiple is by evaluating $Q_n(1)$ by writing

$$Q_n(1) = \lim_{x \to 1} \sqrt{\frac{1+x}{1-x}} \frac{d^n}{dx^n}\left\{ (1-x)^{n+1/2}(1+x)^{n-1/2} \right\}.$$

The only term which contributes in the limit $x \to 1$ is $(1+x)^{n-1/2}$ multiplied by the nth derivative of $(1-x)^{n+1/2}$. Thus

$$Q_n(1) = (n+1/2)(n-1/2)...(3/2)2^n(-1)^n. \qquad (8.5.3)$$

Thus

$$W_n(x) = \frac{(-1)^n}{3.5...(2n-1)} \sqrt{\frac{1+x}{1-x}} \frac{d^n}{dx^n}\left\{ (1-x)^{n+1/2}(1+x)^{n-1/2} \right\}. \qquad (8.5.4)$$

8.6 Explicit Expression

The simplest way to obtain a power series expression for $W_n(x)$ is to use Eq. (8.1.2)

$$W_n(x) = U_n(x) + U_{n-1}(x)$$

$$= \sum_{q=0}^{[n/2]} (-1)^q \frac{(n-q)!}{q!\,(n-2q)!} (2x)^{n-2q} + \sum_{q=0}^{[(n-1)/2]} (-1)^q \frac{(n-q-1)!}{q!\,(n-1-2q)!} (2x)^{n-2q-1}.$$

(8.6.1)

We can express x^n in terms of the Chebyshev polynomials of the fourth kind

$$x^n = \sum_{m=0}^{n} a_m W_m(x).$$

(8.6.2)

The simplest way to evaluate the coefficients a_m is to use the trigonometric representation. Noting that

$$(e^{i\theta} + e^{-i\theta})^n (e^{i\theta/2} - e^{-i\theta/2}) = \sum_{m=0}^{n} \binom{n}{m} \left(e^{(n+1/2-2m)i\theta} - e^{(n-1/2-2m)i\theta} \right).$$

If we replace m by $n-m$ in the second exponent on the right-hand side, we get

$$2^n \cos^n \theta \sin \theta/2 = \sum_{m=0}^{n} \binom{n}{m} \sin[(n-2m+1/2)\theta]$$

and so

$$x^n = 2^{-n} \sum_{m=0}^{[n/2]} \binom{n}{m} W_{n-2m} - 2^{-n} \sum_{m=0}^{[(n-1)/2]} \binom{n}{m} W_{n-1-2m},$$

(8.6.3)

where $[n/2]$ is the largest integer less than or equal to $n/2$.

An alternative way to evaluate the expansion coefficients a_m is to multiply equation (8.6.4) by $W_q(x)$ and integrating from -1 to 1. Making use of the orthogonality property of the Chebyshev polynomials, we see

$$a_q h_q = \int_{-1}^{1} x^n W_q(x) \sqrt{\frac{1-x}{1+x}}\, dx = \frac{(-2)^q\, q!}{(2q)!} \int_{-1}^{1} x^n \frac{d^q}{dx^q} \left[(x^2-1)^q \sqrt{\frac{1-x}{1+x}} \right] dx$$

$$= \frac{2^q\, q!\, n!}{(2q)!\,(n-q)!} \int_{-1}^{1} x^{n-q}(1-x)(1-x^2)^{q-1/2} dx$$

after integrating by parts q times with the integrated parts being zero at the end points. Looking at the factor $(1-x)$, we see that if $n-q$ is even, the contribution of the x part is zero since the integrand with x is odd. Similarly, the contribution of the 1 will be zero if $n - q$ is odd. We therefore split the integral into two parts writing $q = n - 2m$ in one part and $q = n - 1 - 2m$. Then, with $h_q = \pi$

$$\pi a_{n-2m} = \frac{2^{n-2m}\, n!\, (n - 2m)!}{(2n - 4m)!\, (2m)!} \int_{-1}^{1} x^{2m}(1 - x^2)^{n-2m-1/2} dx,$$

$$\pi a_{n-1-2m} = -\frac{2^{n-1-2m}\, n!\, (n - 1 - 2m)!}{(2n - 4m - 2)!\, (2m + 1)!} \int_{-1}^{1} x^{2m+2}(1 - x^2)^{n-2m-3/2} dx.$$

Writing $t = x^2$ converts the first integral into the Beta function

$$B(m + 1/2, n - 2m + 1/2) = \frac{\Gamma(m + 1/2)\Gamma(n - 2m + 1/2)}{\Gamma(n - m + 1)}$$

and the second integral to

$$B(m + 3/2, n - 2m - 1/2) = \frac{\Gamma(m + 3/2)\Gamma(n - 2m - 1/2)}{\Gamma(n - m + 1)}.$$

(See the General Appendix for the properties of the Beta function.) Then

$$x^n = \sum_{m=0}^{[n/2]} \frac{2^{-n}\, n!}{m!\, (n - m)!} W_{n-2m}(x) - \sum_{m=0}^{[(n-1)/2]} \frac{2^{-n}\, n!}{m!\, (n - m)!} W_{n-1-2m}(x),$$

where $[n/2]$ is the largest integer less than or equal to $n/2$.

8.7   Generating Functions

The simplest generating function can be derived by taking the imaginary part of the identity

$$\sum_{n=0}^{\infty} t^n e^{n+1/2)i\theta} = \frac{e^{i\theta/2}}{1 - te^{i\theta}} \qquad |t| < 1 \qquad (8.7.1)$$

and putting $x = \cos\theta$.

This leads to the relation

$$\sum_{n=0}^{\infty} t^n \sin[(n + 1/2)\theta] = \sin(\theta/2)\frac{1 + t}{1 - 2t\cos\theta + t^2}.$$

Dividing through by $\sin(\theta/2)$ and putting $x = \cos\theta$, we have from the trigonometric representation

$$\sum_{n=0}^{\infty} t^n W_n(x) = \frac{1 + t}{1 - 2xt + t^2}. \qquad (8.7.2)$$

The derivation of this generating function is a special case of using the identity

$$\sum_{n=0}^{\infty} \frac{\Gamma(n+\mu)}{n!\,\Gamma(\mu)} t^n e^{i(n+1/2)\theta} = \frac{e^{i\theta/2}}{(1 - te^{i\theta})^\mu} \qquad (8.7.3)$$

and taking the imaginary part.

We can derive other generating functions by choosing different values for μ. If we let $\mu = 2$,

$$\sum_{n=0}^{\infty} (n+1)t^n \sin[(n+1/2)\theta] = Im\left(\frac{e^{i\theta/2}}{(1 - te^{i\theta})^2} \right)$$

$$= \frac{\sin(\theta/2)[1 + 2t - t^2(2x+1)]}{R^4},$$

where $R = \sqrt{1 - 2xt + t^2}$. Therefore

$$\sum_{n=0}^{\infty} (n+1)t^n W_n(x) = \frac{1 + 2t - t^2(2x+1)}{R^4}. \qquad (8.7.4)$$

We can derive another generating function if we let $\mu = 1/2$,

$$\sum_{n=0}^{\infty} \frac{\Gamma(n+1/2)}{n!\,\Gamma(1/2)} t^n \sin[(n+1/2)\theta] = Im\left(\frac{e^{i\theta/2}}{\sqrt{1 - te^{i\theta}}} \right)$$

$$= Im\left[\frac{e^{i\theta/2}}{\sqrt{2R}} \left(\sqrt{1 - xt + R} + \frac{it\sin\theta}{\sqrt{1 - xt + R}} \right) \right] = \frac{\sin(\theta/2)(1 + t + R)}{R\sqrt{2(1 - xt + R)}}.$$

Therefore

$$\sum_{n=0}^{\infty} \frac{\Gamma(n+1/2)}{n!\,\Gamma(1/2)} t^n W_n(x) = \frac{1 + t + R}{R\sqrt{2(1 - xt + R)}}. \qquad (8.7.5)$$

For $\mu = -1/2$,

$$\sum_{n=0}^{\infty} \frac{\Gamma(n-1/2)}{n!\,\Gamma(-1/2)} t^n \sin[(n+1/2)\theta] = Im\left(e^{i\theta/2}\sqrt{1 - te^{i\theta}} \right)$$

$$= Im\left[\frac{e^{i\theta/2}}{\sqrt{2}} \left(\sqrt{1 - xt + R} - \frac{it\sin\theta}{\sqrt{1 - xt + R}} \right) \right] = \frac{\sin(\theta/2)(1 - 2xt - t + R)}{\sqrt{2(1 - xt + R)}}.$$

Therefore

$$\sum_{n=0}^{\infty} \frac{\Gamma(n-1/2)}{n!\,\Gamma(-1/2)} t^n W_n(x) = \frac{1 - 2xt - t + R}{\sqrt{2(1 - xt + R)}}. \qquad (8.7.6)$$

An alternative derivation of the validity of these generating functions is to define functions $\phi_n(x)$

$$\sum_{n=0}^{\infty} t^n \phi_n(x) = w(x,t), \tag{8.7.7}$$

for each function $w(x,t)$ on the right-hand side of (8.7.2), (8.7.4), (8.7.5) or (8.7.6). We can see in each case that $\phi_n(x)$ is an mth order polynomial in x by using the binomial theorem to expand the appropriate $w(x,t)$ in powers of t and collecting the terms in t^n. The next step is to show that for each possible function w,

$$(1-x^2)\frac{\partial^2 w(x,t)}{\partial x^2} - (1+2x)\frac{\partial w(x,t)}{\partial x} = -t\frac{\partial^2 tw(x,t)}{\partial t^2}. \tag{8.7.8}$$

This can be done most easily by using a computer algebra package such as Maple.

If we substitute the left-hand side of (8.7.7) into (8.7.8) then

$$\sum_{n=0}^{\infty} t^n \left\{ (1-x^2)\frac{d^2 \phi_n(x)}{dx^2} - (1+2x)\frac{d\phi_n(x)}{dx} \right\}$$

$$= -t \sum_{n=0}^{\infty} \phi_n(x)\frac{d^2}{dx^2}t^{n+1} = -\sum_{n=0}^{\infty} n(n+1)t^t\phi_n(x). \tag{8.7.9}$$

On equating powers of t on both sides of this equation we see that $\phi_n(x)$ satisfies the Chebyshev differential equation and must therefore be a multiple of the Chebyshev polynomial of the fourth kind, that is $a_n W_n(x)$. We can find the value of a_n for each generating function by putting $x = 1$ in each of the expressions on the right hand side of Eqs. (8.7.2), (8.7.4), (8.7.5) and (8.7.6) and expanding in powers of t.

For generating function (8.7.2), $\mu = 1$,

$$w(1,t) = \frac{1+t}{(1-t)^2} = 1 + 3t + 5t^2 + 7t^3 + \ldots(2n+1)t^n + \ldots$$

and so $a_n = 1$.

For (8.7.4), $\mu = 2$,

$$w(1,t) = \frac{1+2t-3t^2}{(1-t)^4} = 1 + 6t + 15t^2 + 28t^3 + \ldots + (n+1)(2n+1)t^n + \ldots$$

and so $a_n = n+1$.

For (8.7.5), $\mu = 1/2$,

$$w(1,t) = \frac{1}{(1-t)^{3/2}} = 1 + \frac{3}{2}t + \frac{\frac{3}{2}\cdot\frac{5}{2}}{1.2}t^2 + \ldots + \frac{\frac{3}{2}\ldots(n+1/2)}{n!}t^n + \ldots$$

and so $a_n = \Gamma(n + 1/2)/[n!\Gamma(1/2)]$.

For (8.7.6), $\mu = -1/2$,

$$w(1, t) = \frac{1 - 2t}{\sqrt{1 - t}} = 1 - \frac{1}{2}3t - \frac{\frac{1}{2} \cdot \frac{1}{2}}{1.2}5t^2 - \frac{\frac{1}{2} \cdot \frac{1}{2} \cdot \frac{3}{2}}{1.2.3}7t^3 - \frac{\frac{1}{2} \cdot \frac{1}{2} \dots (n - 3/2)}{n!}(2n+1)t^2 - \dots$$

and so $a_n = \Gamma(n - 1/2)/[n!\,\Gamma(-1/2)]$.

8.8 Recurrence Relations

There is a simple recurrence relation between the Chebyshev polynomials which is most easily derived using the trigonometric identity $\sin(A + B) = \sin A \cos B + \cos A \sin B$.

$$W_{n+1}(x) + W_{n-1}(x) = 2xW_n(x). \tag{8.8.1}$$

This relation can be simply derived by writing the generating function Eq. (8.7.2) in the form

$$(1 - 2xt + t^2) \sum_{n=0}^{\infty} t^n W_n(x) = 1 + t.$$

A more general recurrence relation is

$$W_{n+m}(x) + W_{n-m}(x) = 2W_n(x)T_m(x), \tag{8.8.2}$$

where $T_m(x)$ is a Chebyshev polynomial of the first kind.

8.9 Differential Relation

We can derive a relation for the derivative W_n' by substituting $x = \cos\theta$

$$(1-x^2)W_n'(x) = \frac{1}{2}\left(n+\frac{1}{2}\right)\left(W_{n-1}(x)-W_{n+1}(x)\right)+\frac{1}{2}(1+x)W_n(x). \tag{8.9.1}$$

8.10 Step Up and Step Down Operators

These operators can be easily derived from the relations (8.8.1) and (8.9.1).

$$W_{n-1}(x) = S_n^- W_n(x) = \frac{2}{2n+1}\left\{nx - 1/2 + (1 - x^2)\frac{d}{dx}\right\}W_n(x) \tag{8.10.1}$$

and

$$W_{n+1}(x) = S_n^+ W_n(x) = \frac{2}{2n+1}\left\{(n+1)x + 1/2 - (1 - x^2)\frac{d}{dx}\right\}W_n(x). \tag{8.10.2}$$

References

Doman B G S, International Journal of Pure and Applied Mathematics, 2010, 63, pp. 197-205.

Ferrar W L, A Text-book of Convergence, Oxford University Press, 1938.

Hochstrasser Urs W, Orthogonal Polynomials, Chapter 22, Handbook of Mathematical Functions, Eds. Abramowitz M and Stegun I A, Dover, 1970.

Koornwinder T H, Wong R, Koekoek R and Swarttouw R F, Chapter 18, NIST Handbook of Mathematical Functions, Eds. Olver W J, Lozier D W, Boisvert R F and Clark C W, NIST and Cambridge University Press, 2009.

Mason J C and Handscomb D C, Chebyshev Polynomials, Chapman & Hall, 2002.

Phillips E G, A Course of Analysis, Cambridge University Press, 1930.

Chapter 9

Gegenbauer Polynomials

9.1 Introduction

Gegenbauer polynomials or ultra spherical polynomials $C_n^{(\alpha)}(x)$ can be obtained using the Gram-Schmidt orthogonalisation process for polynomials in the domain $(-1,1)$ with the weight factor $(1-x^2)^{\alpha-1/2}$, $\alpha > -1/2$. $C_n^{(0)}(x)$ is defined as $\lim_{\alpha\to 0} C_n^{(\alpha)}(x)/\alpha$. For $\alpha \neq 0$ the resulting polynomial $R_n(x)$ is multiplied by a number which makes the value at $x = 1$ equal to $(2\alpha)_n/n! = 2\alpha(2\alpha+1)(2\alpha+2)...(2\alpha+n-1)/n!$. For $\alpha = 0$ and $n \neq 0$ the value at $x = 1$ is $2/n$ while $C_0^{(0)}(x) = 1$.

The Gegenbauer polynomials are multiples of the Jacobi polynomials $P_n^{(\alpha-1/2,\alpha-1/2)}(x)$. In fact

$$C_n^{(\alpha)}(x) = \frac{\Gamma(2\alpha+n)\Gamma(\alpha+1/2)}{\Gamma(2\alpha)\Gamma(\alpha+n+1/2)} P_n^{(\alpha-1/2,\alpha-1/2)}(x). \tag{9.1.1}$$

The Legendre polynomials $P_n(x)$ are equal to $C_n^{(1/2)}(x)$. The Gegenbauer polynomial $C_n^{(0)}(x) = 2T_n(x)/n$, where $T_n(x)$ is the Chebyshev polynomial of the first kind and $C_n^{(1)}(x) = U_n(x)$, the Chebyshev polynomial of the second kind.

The first few Gegenbauer polynomials of order α are:

$C_0^{(\alpha)}(x) = 1.$
$C_1^{(\alpha)}(x) = 2\alpha x.$
$C_2^{(\alpha)}(x) = \alpha_2\, 2x^2 - \alpha.$
$C_3^{(\alpha)}(x) = \alpha_3\, 4x^3/3 - 2\alpha_2 x.$
$C_4^{(\alpha)}(x) = \alpha_4\, 2x^4/3 - \alpha_3\, 2x^2 + \alpha_2.$
$C_5^{(\alpha)}(x) = \alpha_5\, 4x^5/15 - \alpha_4\, 4x^3/3 + \alpha_3 x.$
$C_6^{(\alpha)}(x) = \alpha_6\, 4x^6/45 - \alpha_5\, 2x^4/3 + \alpha_4\, x^2/2 - \alpha_3,$
where the Pochammer symbol $a_n = a(a+1)...(a+n-1)$.

Graphs of $C_n^{(2)}(x)$ for $n = 1$ to $n = 4$.

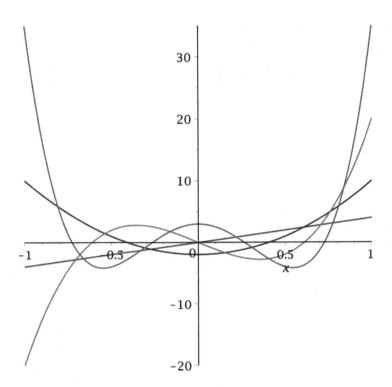

9.2 Differential Equation

The ordinary differential equation satisfied by Gegenbauer polynomials is

$$(1 - x^2)\frac{d^2y}{dx^2} - (2\alpha + 1)x\frac{dy}{dx} + \lambda y = 0. \tag{9.2.1}$$

The point $x = 0$ is an ordinary point. This means that we can express the solution in the form of a power series $y = \sum_{n=0}^{\infty} a_n x^n$.

Substituting this into the differential equation leads to

$$(1 - x^2)\sum_{n=0}^{\infty} n(n-1)a_n x^{n-2} - (2\alpha + 1)x\sum_{n=0}^{\infty} na_n x^{n-1} + \lambda \sum_{n=0}^{\infty} a_n x^n = 0.$$

If we equate the coefficient of x^n to zero in this equation, we get

$$(n+2)(n+1)a_{n+2} - n(n-1)a_n - (2\alpha+1)na_n + \lambda a_n = 0.$$

This leads to the recurrence relation

$$\frac{a_{n+2}}{a_n} = \frac{n(n+2\alpha) - \lambda}{(n+2)(n+1)}. \tag{9.2.2}$$

We see that there are two power series solutions, one series containing only even powers of x and the other an odd power series.

The ratio test can be used to show that both series converge for $|x| < 1$, but the test fails for $|x| = 1$. We need a more powerful test to determine whether the series converge for $|x| = 1$. If we let $u_m = a_{2m}$, then

$$\frac{u_m}{u_{m+1}} = \frac{(2m+2)(2m+1)}{2m(2m+2\alpha) - \lambda} = 1 + \frac{1.5 - \alpha}{m} + O\left(\frac{1}{m^2}\right).$$

The coefficient of $1/m$ is $1.5 - \alpha$ and so using Gauss's test (see e.g. Philips, p. 129 or Ferrar, p. 124), the even series converges for $\alpha < 0.5$ and diverges for $\alpha \geq 0.5$. By the same argument, the odd series also converges for $\alpha < 0.5$ and diverges for $\alpha \geq 0.5$.

If we examine the recurrence relation, we see that if $\lambda = m(m + 2\alpha)$, where m is an integer, one of the series will terminate but the other is an infinite series. If this integer m is an even integer, the terminating series will be an even polynomial, whereas if m is odd, the polynomial will be odd. Let us for the moment denote the polynomial by $R_m(x)$. We are now going to show that these $R_m(x)$ obey the same orthogonality relations as the Gegenbauer polynomials and must therefore be multiples of them.

9.3 Orthogonality

The differential equation (1.1) for $R_m(x)$ can be written in Sturm Liouville form

$$\frac{d}{dx}\left[(1-x^2)^{\alpha+1/2}\frac{dR_m(x)}{dx}\right] = -m(m+2\alpha)(1-x^2)^{\alpha-1/2}R_m(x). \tag{9.3.1}$$

If we multiply this by $R_n(x)$ and integrate from -1 to 1, the left-hand side becomes

$$\int_{-1}^{1} R_n(x)\frac{d}{dx}\left[(1-x^2)^{\alpha+1/2}R'_m(x)\right]dx.$$

On integrating this by parts and noting that if $\alpha > -1/2$, the integrated term vanishes at both end points, we find that

$$\int_{-1}^{1} (1 - x^2)^{\alpha+1/2} R'_n(x) R'_m(x) dx$$

$$= m(m + 2\alpha) \int_{-1}^{1} (1 - x^2)^{\alpha-1/2} R_n(x) R_m(x) dx.$$

If we follow the same procedure as above but with m and n interchanged we will produce an equation which is the same as that above on the left-hand side and the same integral on the right but with a coefficient $n(n + 2\alpha)$ instead of $m(m + 2\alpha)$. If we subtract one of these equations from the other we get

$$[m(m + 2\alpha) - n(n + 2\alpha)] \int_{-1}^{1} (1 - x^2)^{\alpha-1/2} R_n(x) R_m(x) dx = 0.$$

Thus for $n \neq m$

$$\int_{-1}^{1} (1 - x^2)^{\alpha-1/2} R_n(x) R_m(x) dx = 0. \qquad (9.3.2)$$

The polynomials $R_m(x)$ thus satisfy the same orthogonality relations as the Gegenbauer polynomials and therefore must be multiples of them. If we multiply $R_n(x)$ by a constant so that $R_n(1) = 2/n$ if $\alpha = 0$ or $R_n(1) = 2\alpha(2\alpha + 1)(2\alpha + 2)...(2\alpha + n - 1)/n!$ if $\alpha \neq 0$, we will produce the Gegenbauer polynomial.

Since x^n can be represented as a linear combination of the Gegenbauer polynomials $C_p^{(\alpha)}$ for $0 \leq p \leq n$, we can deduce that

$$\int_{-1}^{1} (1 - x^2)^{\alpha-1/2} x^n C_m^{(\alpha)}(x) dx = 0 \qquad \text{for all } n < m. \qquad (9.3.3)$$

9.4   Rodrigues Formula

By treating $(1 - x^2)^{n+\alpha-1/2}$ as the product $(1 - x)^{n+\alpha-1/2}(1 + x)^{n+\alpha-1/2}$ and using the formula for differentiating the product uv m times, we see that if $n > m$ then

$$\frac{d^m}{dx^m}(1 - x^2)^{n+\alpha-1/2} = \frac{d^m}{dx^m}\left[(1 - x)^{n+\alpha-1/2}(1 + x)^{n+\alpha-1/2}\right]$$

contains a factor $(1 - x^2)^{n-m+\alpha-1/2}$ which is zero when $x = \pm 1$ (note that $\alpha > -1/2$).

From this result, we see that

$$\int_{-1}^{1} \frac{d^n}{dx^n}(1-x^2)^{n+\alpha-1/2}dx = 0.$$

If we integrate by parts we see that

$$\int_{-1}^{1} x\frac{d^n}{dx^n}(1-x^2)^{n+\alpha-1/2}dx = 0 \qquad \text{provided that } n > 1,$$

and integrating by parts m times

$$\int_{-1}^{1} x^m\frac{d^n}{dx^n}(1-x^2)^{n+\alpha-1/2} = 0 \qquad \text{provided that } n > m. \qquad (9.4.1)$$

Let us define the nth order polynomial $Q_n(x)$ by

$$Q_n(x) = \frac{1}{(1-x^2)^{\alpha-1/2}}\frac{d^n}{dx^n}\left[(1-x^2)^{n+\alpha-1/2}\right]. \qquad (9.4.2)$$

Then

$$\int_{-1}^{1}(1-x^2)^{\alpha-1/2}x^m Q_n(x)\,dx = 0 \qquad m < n$$

is orthogonal to x^m for all values of $m < n$. This means that $Q_n(x)$ is orthogonal to $Q_m(x)$ for all values of $m < n$. These polynomials must therefore be multiples of the Gegenbauer polynomials we found earlier. We can find out what this multiple is by evaluating $Q_n(1)$.

$$Q_n(1) = \lim_{x \to 1}\left\{\frac{1}{(1-x^2)^{\alpha-1/2}}\frac{d^n}{dx^n}\left[(1-x)^{n+\alpha-1/2}(1+x)^{n+\alpha-1/2}\right]\right\}.$$

The only term which contributes in the limit $x \to 1$ is the nth derivative of $(1-x)^{n+\alpha-1/2}$ multiplied by $(1+x)^{n+\alpha-1/2}$. Thus

$$Q_n(1) = (n+\alpha-1/2)(n+\alpha-3/2)...(\alpha+1/2)2^n(-1)^n.$$

Therefore

$$C_n^{(\alpha)}(x) = \frac{(2\alpha)(2\alpha+1)...(2\alpha+n-1)}{2^n n!(\alpha+1/2)...(n+\alpha-1/2)}\frac{(-1)^n}{(1-x^2)^{\alpha-1/2}}\frac{d^n}{dx^n}(1-x^2)^{n+\alpha-1/2}$$

$$= \frac{\Gamma(2\alpha+n)\Gamma(\alpha+1/2)}{2^n n!\Gamma(2\alpha)\Gamma(\alpha+n+1/2)}\frac{(-1)^n}{(1-x^2)^{\alpha-1/2}}\frac{d^n}{dx^n}(1-x^2)^{n+\alpha-1/2}. \qquad (9.4.3)$$

We can now evaluate

$$h_n = \int_{-1}^{1}(1-x^2)^{\alpha-1/2}[C_n^{(\alpha)}(x)]^2 dx.$$

Using the formula (9.4.3),

$$h_n = \frac{(-1)^n \Gamma(2\alpha + n)\Gamma(\alpha + 1/2)}{2^n n! \Gamma(2\alpha)\Gamma(\alpha + n + 1/2)} \int_{-1}^{1} C_n^{(\alpha)}(x) \frac{d^n}{dx^n} (1 - x^2)^{n+\alpha-1/2} dx.$$

Integrating by parts n times and using $d^m/dx^m (1 - x^2)^{n+\alpha-1/2} = 0$ at $x = \pm 1$ for $m < n$

$$h_n = \frac{\Gamma(2\alpha + n)\Gamma(\alpha + 1/2)}{2^n n! \Gamma(2\alpha)\Gamma(\alpha + n + 1/2)} \int_{-1}^{1} (1 - x^2)^{n+\alpha-1/2} \frac{d^n}{dx^n} C_n^{(\alpha)}(x) dx$$

$$= \frac{\Gamma(2\alpha + n)\Gamma(\alpha + 1/2)\Gamma(\alpha + n)}{n! \Gamma(2\alpha)\Gamma(\alpha + n + 1/2)\Gamma(\alpha)} \int_{-1}^{1} (1 - x^2)^{n+\alpha-1/2} dx,$$

where we have used the expression for k_n, the coefficient of x^n in $C_n^{(\alpha)}(x)$ derived in the next section 9.5. Writing $t = x^2$ gives

$$h_n = \frac{\Gamma(2\alpha + n)\Gamma(\alpha + 1/2)\Gamma(\alpha + n)}{n! \Gamma(2\alpha)\Gamma(\alpha + n + 1/2)\Gamma(\alpha)} \int_{0}^{1} t^{-1/2} (1 - t)^{n+\alpha-1/2} dt,$$

$$h_n = \frac{\Gamma(2\alpha + n)\Gamma(\alpha + 1/2)\Gamma(\alpha + n)}{n! \Gamma(2\alpha)\Gamma(\alpha + n + 1/2)\Gamma(\alpha)} B(1/2, n + \alpha + 1/2),$$

$$h_n = \frac{\Gamma(2\alpha + n)\Gamma(\alpha + 1/2)\Gamma(\alpha + n)}{n! \Gamma(2\alpha)\Gamma(\alpha + n + 1/2)\Gamma(\alpha)} \frac{\Gamma(1/2)\Gamma(n + \alpha + 1/2)}{\Gamma(n + \alpha + 1)},$$

$$h_n = \frac{2^{1-2\alpha} \pi \Gamma(n + 2\alpha)}{n!(n + \alpha)[\Gamma(\alpha)]^2}. \tag{9.4.4}$$

We have used the expression for the Beta function $B(\alpha, \beta) = \Gamma(\alpha)\Gamma(\beta)/\Gamma(\alpha + \beta)$ and the duplication formula for Gamma functions $\Gamma(2z) = 2^{2z-1}\Gamma(z)\Gamma(z + 1/2)/\sqrt{\pi}$. Properties of the Gamma function and the Beta function are described in the General Appendix.

For $C_n^{(0)}(x)$, $h_n = 2\pi/n^2$ for $n > 0$.

9.5 Explicit Expression

We use the recurrence relation for coefficients of successive terms of the polynomial $C_n^{(\alpha)}$ obtained in the solution of the differential equation, (9.2.2) with $\lambda = n(n + 2\alpha)$

$$\frac{a_m}{a_{m-2}} = \frac{(m - n - 2)(m - 2 + n + 2\alpha)}{m(m - 1)},$$

and the coefficient of x^n, $k_n = 2^n \alpha_n / n!$ to obtain the expression for $C_n^{(\alpha)}$ as a polynomial.

$$C_n^{(\alpha)}(x) = \frac{1}{\Gamma(\alpha)} \sum_{m=0}^{[n/2]} \frac{(-1)^m \Gamma(\alpha + n - m)}{m!(n-2m)!} (2x)^{(n-2m)}, \tag{9.5.1}$$

where $[n/2]$ is the largest integer $\leq n/2$.

The coefficient of x^{n-1}, k_n' is zero since there are only even powers of x if n is even and only odd powers if n is odd.

For $\alpha = 0$, $k_n = 2^n/n$

$$C_n^{(0)}(x) = \sum_{m=0}^{[n/2]} \frac{(-1)^m (n-m-1)!}{m!(n-2m)!} (2x)^{n-2m}. \tag{9.5.2}$$

We can express x^n in terms of the Gegenbauer polynomials

$$x^n = \sum_{m=0}^{n} a_m C_m^{(\alpha)}(x).$$

The coefficients a_m can be calculated by multiplying this equation by $C_q^{(\alpha)}(x)$ and integrating from -1 to 1. Making use of the orthogonality property of the Gegenbauer polynomials, we see

$$a_q = \frac{1}{h_q} \int_{-1}^{1} x^n C_q^{(\alpha)}(x) dx = (-1)^q A_q \int_{-1}^{1} x^n \frac{d^q}{dx^q} (1-x^2)^{q+\alpha-1/2} dx$$

where we have used the Rodrigues formula Eq. (9.4.3) and

$$A_q = \frac{(q+\alpha)\Gamma(\alpha)}{2^q \sqrt{\pi} \, \Gamma(\alpha+q+1/2)}.$$

Then

$$a_q = \frac{A_q n!}{(n-q)!} \int_{-1}^{1} x^{n-q} (1-x^2)^{q+\alpha-1/2} dx$$

after integrating by parts q times with the integrated parts being zero at the end points. If $n - q$ is odd, this integral will be zero. Therefore let $n - q = 2m$, then

$$a_{n-2m} = \frac{n! A_{n-2m}}{(2m)!} \int_{-1}^{1} x^{2m} (1-x^2)^{n-2m+\alpha-1/2} dx.$$

Writing $t = x^2$ converts the integral into the Beta function

$$B(m+1/2, n-2m+\alpha+1/2) = \frac{\Gamma(m+1/2)\Gamma(n-2m+\alpha+1/2)}{\Gamma(n-m+\alpha+1)}$$

(see the General Appendix for the properties of the Beta function). Then

$$x^n = n! \sum_{m=0}^{[n/2]} \frac{(n-2m+\alpha)\Gamma(\alpha)}{2^n \, m! \, \Gamma(n-m+\alpha+1)} C_{n-2m}^{(\alpha)}(x), \tag{9.5.3}$$

where $[n/2]$ is the largest integer less than or equal to $n/2$.

9.6   Generating Function

For $\alpha \neq 0$, let

$$w(x,t) = \frac{1}{(1 - 2xt + t^2)^\alpha}. \tag{9.6.1}$$

It can be shown that

$$(1 - x^2)\frac{\partial^2 w(x,t)}{\partial x^2} - (2\alpha + 1)x\frac{\partial w(x,t)}{\partial x} = -t^{1-2\alpha}\frac{\partial}{\partial t}\left\{t^{2\alpha+1}\frac{\partial w(x,t)}{\partial t}\right\}. \tag{9.6.2}$$

This can conveniently be done using a computer algebra package such as Maple. Let us define $\phi_n(x)$ by

$$\sum_{n=0}^{\infty} t^n \phi_n(x) = \frac{1}{(1 - 2xt + t^2)^\alpha} \qquad |t| < 1. \tag{9.6.3}$$

On expanding the right-hand side using the binomial theorem and collecting all of the terms in t^n, we see that the functions $\phi_n(x)$ are nth order polynomials in x.

Substituting the left-hand side of (9.6.3) into (9.6.2) leads to

$$\sum_{n=0}^{\infty} t^n \left\{(1 - x^2)\frac{d^2\phi_n(x)}{dx^2} - (2\alpha + 1)x\frac{d\phi_n(x)}{dx}\right\}$$

$$= -t^{1-2\alpha}\sum_{n=0}^{\infty} \phi_n(x)\frac{d}{dt}\left\{t^{2\alpha+1}\frac{d}{dt}t^n\right\} = -\sum_{n=0}^{\infty} n(n + 2\alpha)t^n\phi_n(x). \tag{9.6.4}$$

If we now equate the coefficients of powers of t on both sides of the equation,

$$(1 - x^2)\frac{d^2\phi_n(x)}{dx^2} - (2\alpha + 1)x\frac{d\phi_n(x)}{dx} = -n(n + 2\alpha)\phi_n(x). \tag{9.6.5}$$

In other words $\phi_n(x)$ satisfies Gegenbauer's equation and is therefore some multiple of the Gegenbauer polynomial $C_n^{(\alpha)}(x)$.

If we put $x = 1$ in Eq. (9.6.3)

$$\sum_{n=0}^{\infty} t^n \phi_n(1) = \frac{1}{(1 - t)^{2\alpha}} = 1 + 2\alpha t + \frac{2\alpha(2\alpha + 1)}{2}t^2 + \frac{2\alpha(2\alpha + 1)(2\alpha + 2)}{3!}t^3 + \dots .$$

Therefore

$$\phi_n(1) = 2\alpha(2\alpha + 1)(2\alpha + 2)\dots(2\alpha + n - 1)/n!$$

so that $C_n^{(\alpha)}(x) = \phi_n(x)$, the nth Gegenbauer polynomial. Then

$$\sum_{n=0}^{\infty} t^n C_n^{(\alpha)}(x) = \frac{1}{(1 - 2xt + t^2)^\alpha} \qquad |t| < 1. \tag{9.6.6}$$

Another generating function can be obtained by writing $\beta = \alpha$ in the generating function for the Jacobi polynomials (see Chap. 11). Thus if $w(x,t)$ is defined as

$$w(x,t) = \frac{2^{\alpha-1/2}}{R(1 + R - xt)^{\alpha-1/2}}, \quad \text{where} \quad R = \sqrt{1 - 2xt + t^2}, \quad (9.6.7)$$

it can be shown that $w(x,t)$ satisfies Eq. (9.6.2). Therefore if $w(x,t)$ is written in the form $w(x,t) = \sum_{n=0}^{\infty} t^n \psi_n(x)$, the functions $\psi_n(x)$ satisfy Eq. (9.6.4) and are therefore multiples of the Chebyshev polynomials $C_n^{(\alpha)}(x)$. We can work out what this multiple is by evaluating

$$w(1,t) = \frac{1}{(1-t)^{\alpha+1/2}} = \sum_{n=0}^{\infty} \psi_n(1)t^n = \sum_{n=0}^{\infty} (\alpha + 1/2)_n t^n/n!.$$

Therefore $\psi_n(x) = (\alpha + 1/2)_n C_n^{(\alpha)}(x)/(2\alpha)_n$.

If we write the expansion of the generating function Eq. (9.6.6) in the form

$$\frac{1}{(1 - 2xt + t^2)^\alpha} = \sum_{n=0}^{\infty} \alpha_n \frac{[t(2x-t)]^n}{n!},$$

we see that k_n the coefficient of x^n in $C_n^{(\alpha)}(x)$ is $2^n \alpha_n/n!$.

For $C_n^{(0)}(x)$ there are a number of generating functions. The chapter on Chebyshev polynomials of the first kind contains a number of such generating functions. For $C_n^{(0)}(x)$ we can use the relation $C_n^{(0)}(x) = 2T_n(x)/n$.

There are also a number of generating functions for $C_n^{(1)}(x) = U_n(x)$, the Chebyshev polynomial of the second kind. These are detailed in the chapter on Chebyshev polynomials of the second kind.

9.7 Recurrence Relations

We can use the generating function to derive a number of recurrence relations. Firstly, if we differentiate Eq. (9.6.6) with respect to t, we find

$$\frac{\partial}{\partial t} \frac{1}{(1 - 2x + t^2)^\alpha} = \frac{\alpha(2x - 2t)}{(1 - 2x + t^2)^{\alpha+1}}. \quad (9.7.1)$$

Substituting the left-hand side of (9.6.6) leads to

$$\sum_{n=1}^{\infty} n t^{n-1} C_n^{(\alpha)}(x) = 2\alpha(x - t) \sum_{n=0}^{\infty} t^n C_n^{(\alpha+1)}(x).$$

Equating the coefficients of t^n leads to the recurrence relation

$$(n+1)C_{n+1}^{(\alpha)}(x) = 2\alpha x C_n^{(\alpha+1)}(x) - 2\alpha C_{n-1}^{(\alpha+1)}. \qquad (9.7.2)$$

Multiplying (9.7.1) by $(1 - 2xt + t^2)$ leads to the equation:

$$(1 - 2xt + t^2) \sum_{n=0}^{\infty} n t^{n-1} C_n^{(\alpha)}(x) = 2\alpha(x - t) \sum_{n=0}^{\infty} t^n C_n^{(\alpha)}(x).$$

If we now equate the coefficients of t^n on both sides of the equation we see that

$$(n+1)C_{n+1}^{(\alpha)}(x) = 2(n+\alpha)x C_n^{(\alpha)}(x) - (n + 2\alpha - 1)C_{n-1}^{(\alpha)}(x). \qquad (9.7.3)$$

We can obtain another recurrence relation from the identity

$$\left(t\frac{\partial}{\partial t} + \alpha\right) \frac{1}{(1 - 2xt + t^2)^\alpha} = \frac{\alpha(1 - t^2)}{(1 - 2xt + t^2)^{\alpha+1}}.$$

If we write this equation in terms of the expansion in powers of t and then equate the coefficients of t^n on both sides, we get

$$(n+\alpha)C_n^{(\alpha)} = \alpha\left(C_n^{(\alpha+1)}(x) - C_{n-2}^{(\alpha+1)}\right). \qquad (9.7.4)$$

Another relation comes from the identity:

$$\frac{2\alpha(1 - x^2)}{(1 - 2xt + t^2)^{\alpha+1}} - t^{1-2\alpha}\frac{\partial}{\partial t}\left\{\frac{t^{2\alpha}}{(1 - 2xt + t^2)^\alpha}\right\} + x\frac{\partial}{\partial t}\left\{\frac{1}{(1 - 2xt + t^2)^\alpha}\right\} = 0.$$

Again, if we substitute the left-hand side of (9.7.6) and equate the coefficients of t^n to zero,

$$2\alpha(1 - x^2)C_n^{(\alpha+1)}(x) = (n + 2\alpha)C_n^{(\alpha)}(x) - (n+1)x C_{n+1}^{(\alpha)}(x). \qquad (9.7.5)$$

And furthermore

$$t^{2-2\alpha}\frac{\partial}{\partial t}\left\{t^{2\alpha-1}\frac{\partial}{\partial t}\left(\frac{1}{(1 - 2xt + t^2)^{\alpha-1}}\right)\right\}$$
$$= \frac{2(\alpha-1)(2\alpha-1)x}{(1 - 2xt + t^2)^\alpha} - \frac{4\alpha(\alpha-1)(1 - x^2)t}{(1 - 2xt + t^2)^{\alpha+1}}$$

leads to the relation

$$2(\alpha-1)(2\alpha-1)x C_n^{(\alpha)}(x)$$
$$= 4\alpha(\alpha-1)(1 - x^2)C_{n-1}^{(\alpha+1)} + (n + 2\alpha - 1)(n+1)C_{n+1}^{(\alpha-1)}. \qquad (9.7.6)$$

And if in the identity

$$4\alpha(1 - x^2)t^{-\alpha}\frac{\partial}{\partial t}\left\{\frac{t^{\alpha+1}}{(1 - 2xt + t^2)^{\alpha+1}}\right\} + \frac{\partial^2}{\partial t^2}\left\{\frac{1}{1 - 2xt + t^2)^\alpha}\right\}$$
$$- t^{1-2\alpha}\frac{\partial^2}{\partial t^2}\left\{\frac{t^{2\alpha+1}}{(1 - 2xt + t^2)^\alpha}\right\} = 0,$$

we equate the sum of the coefficients of t^n to zero we find

$$4\alpha(n+\alpha+1)(1 - x^2)C_n^{(\alpha+1)}(x)$$
$$= (n + 2\alpha)(n + 2\alpha + 1)C_n^{(\alpha)}(x) - (n+1)(n+2)C_{n+2}^{(\alpha)}(x). \qquad (9.7.7)$$

9.8 Addition Formulae

The relation

$$C_n^{(\alpha+\beta)}(x) = \sum_{m=0}^{n} C_m^{(\alpha)}(x) C_{n-m}^{(\beta)}(x) \tag{9.8.1}$$

can be derived from the generating function as follows:

$$\sum_{n=0}^{\infty} t^n C_n^{\alpha+\beta)}(x) = \frac{1}{(1-2xt+t^2)^{\alpha+\beta}} = \frac{1}{(1-2xt+t^2)^\alpha} \frac{1}{(1-2xt+t^2)^\beta}$$

$$= \sum_{p=0}^{\infty} t^p C_p^{(\alpha)}(x) \sum_{q=0}^{\infty} t^q C_q^{(\alpha)}(x).$$

We get Eq. (9.8.1) by equating the coefficients of t^n on both sides of this equation.

The generating function can be used to show that

$$C_n^{(\alpha+1)} = \sum_{m=0}^{n} \frac{m+2\alpha}{2\alpha} C_m^{(\alpha)}(x) x^{n-m}. \tag{9.8.2}$$

If we multiply the right-hand side by t^n and sum from $n = 0$ to ∞, we obtain

$$\sum_{n=0}^{\infty} t^n \sum_{m=0}^{n} \frac{m+2\alpha}{2\alpha} C_m^{(\alpha)}(x) x^{n-m} = \sum_{m=0}^{\infty} \frac{m+2\alpha}{2\alpha} C_m^{(\alpha)}(x) \sum_{n=m}^{\infty} t^n x^{n-m}$$

$$= \sum_{m=0}^{\infty} \frac{m+2\alpha}{2\alpha} C_m^{(\alpha)}(x) \frac{t^m}{1-xt} = \frac{1}{1-xt} \left(1 + \frac{t}{2\alpha} \frac{\partial}{\partial t} \right) \frac{1}{(1-2xt+t^2)^\alpha}$$

$$= \frac{1}{1-xt} \frac{1-2xt+t^2+t(x-t)}{(1-2xt+t^2)^{\alpha+1}} = \sum_{n=0}^{\infty} t^n C_n^{(\alpha+1)}(x).$$

Equating the coefficients of t^n on both sides gives us Eq. (9.8.2).

9.9 Differential Relations

We can show that

$$(1-x^2)\frac{\partial}{\partial x}\left\{ \frac{1}{(1-2xt+t^2)^\alpha} \right\} - t\frac{\partial}{\partial t}\left\{ \frac{t-x}{(1-2xt+t^2)^\alpha} \right\} = \frac{(2\alpha-1)t}{(1-2xt+t^2)^\alpha}.$$

If we equate coefficients of t^n on both sides of this equation we see that

$$(1-x^2)\frac{d}{dx}C_n^{(\alpha)}(x) = -nx C_n^{(\alpha)}(x) + (n+2\alpha-1)C_{n-1}^{(\alpha)}(x). \tag{9.9.1}$$

If we differentiate the generating function Eq. (9.6.6) with respect to x, we can derive a relation between the derivative of $C_n^{(\alpha)}(x)$ and $C_{n-1}^{(\alpha+1)}(x)$.

$$\sum_{n=0}^{\infty} t^n \frac{d}{dx} C_n^{(\alpha)}(x) = \frac{\partial}{\partial x} \left\{ \frac{1}{(1 - 2xt + t^2)^\alpha} \right\}$$

$$= \frac{2\alpha t}{(1 - 2xt + t^2)^{\alpha+1}} = 2\alpha t \sum_{n=0}^{\infty} t^n C_n^{(\alpha+1)}(x).$$

If we equate powers of t on both sides, we find

$$\frac{d}{dx} C_n^{(\alpha)}(x) = 2\alpha C_{n-1}^{(\alpha+1)}(x). \qquad (9.9.2)$$

If we now differentiate the generating function Eq. (9.6.6) p times

$$\sum_{n=0}^{\infty} t^n \frac{d^p}{dx^p} C_n^{(\alpha)}(x) = \frac{\partial^p}{\partial x^p} \left\{ \frac{1}{(1 - 2xt + t^2)^\alpha} \right\}$$

$$= \frac{2^p (\alpha)_p\, t^p}{(1 - 2xt + t^2)^{\alpha+p}} = 2^p (\alpha)_p\, t^p \sum_{n=0}^{\infty} t^n C_n^{(\alpha+p)}(x).$$

Equating powers of t on both sides leads to

$$\frac{d^p}{dx^p} C_n^{(\alpha)}(x) = 2^p (\alpha)_p\, C_{n-p}^{(\alpha+p)}(x). \qquad (9.9.3)$$

In this equation we have used the Pochammer symbol

$$(\alpha)_p = \alpha(\alpha + 1)...(\alpha + p - 1).$$

This result can be deduced directly from the differential equation Eq. (9.2.1). If we differentiate Eq. (9.2.1) for the polynomial solution p times, we get

$$(1 - x^2)\frac{d^2 z}{dx^2} - (2\alpha + 2p + 1)x\frac{dz}{dx} + (m - p)(m + p + 2\alpha)z = 0.$$

This equation is clearly satisfied by $C_{m-p}^{(\alpha+p)}$ and $d^p C_m^{(\alpha)}(x)/dx^p$. Therefore since the polynomial solution is unique apart from a multiplicative constant, $d^p C_m^{(\alpha)}/dx^p = k C_{m-p}^{(\alpha+p)}(x)$. If we equate the coefficients of x^{m-p} in the two expressions, we find the same result as Eq. (9.8.3). For $C_m^{(0)}(x)$ we get

$$\frac{d^p}{dx^p} C_m^{(0)}(x) = 2^p (p - 1)!\, C_{m-p}^{(p)}(x). \qquad (9.9.4)$$

9.10 Step Up and Step Down Operators

These operators can be easily derived from relations (9.9.1) and (9.7.3).

$$C_{n-1}^{(\alpha)}(x) = S_n^{(\alpha)-} C_n^{(\alpha)}(x) = \frac{1}{n + 2\alpha - 1}\left\{ nx + (1 - x^2)\frac{d}{dx} \right\}C_n^{(\alpha)}(x)$$

(9.10.1)

and

$$C_{n+1}^{(\alpha)}(x) = S_n^{(\alpha)+} C_n^{(\alpha)}(x) = \frac{1}{n + 1}\left\{ (n + 2\alpha)x - (1 - x^2)\frac{d}{dx} \right\}C_n^{(\alpha)}(x).$$

(9.10.2)

References

Courant R and Hilbert D, Methods of Mathematical Physics Vol 1, Interscience Publishers, 1953.

Erdelyi A, Higher Transcendental Functions Vol 2, McGraw-Hill 1953.

Ferrar W L, Textbook of Convergence, Oxford University Press 1938.

Hochstrasser Urs W, Orthogonal polynomials, Chapter 22, Handbook of Mathematical Functions, Eds. Abramowitz M and Stegun I A, Dover, 1970.

Koornwinder T H, Wong R, Koekoek R and Swarttouw R F, Chapter 18, NIST Handbook of Mathematical Functions, Eds. Olver W J, Lozier D W, Boisvert R F and Clark C W, NIST and Cambridge University Press, 2009.

Phillips E G, A Course of Analysis, Cambridge University Press, 1930.

Chapter 10

Associated Legendre Functions

10.1   Introduction

The associated Legendre functions $P_n^m(x)$, where n and m are both integers, arise in many problems involving the solution of Laplace's equation and similar equations in spherical polar coordinates. They are closely related to the Legendre polynomials $P_n(x)$ and the Gegenbauer polynomials $C_{n-m}^{(m+1/2)}(x)$ which are themselves multiples of the Jacobi polynomials $P_{n-m}^{(m,m)}(x)$. In fact for non-negative $m \leq n$.

$$P_n^m(x) = (-1)^m(1-x^2)^{m/2}\frac{d^m}{dx^m}P_n(x) = \frac{(-1)^m(1-x^2)^{m/2}}{2^n n!}\frac{d^{n+m}}{dx^{n+m}}(x^2-1)^n$$

$$= \frac{(-1)^m(2m)!}{2^m m!}(1-x^2)^{m/2}C_{n-m}^{(m+1/2)}(x)$$

$$= (-1)^m\frac{(m+n)!}{2^m n!}(1-x^2)^{m/2}P_{n-m}^{(m,m)}(x). \qquad (10.1.1)$$

These functions are polynomials of order n if m is an even number and $\sqrt{1-x^2}$ times a polynomial of order $n-1$ if m is odd. For $m=0$, they are the Legendre polynomials $P_n(x)$. The factor $(-1)^m$ is known as the Condon-Shortley phase which is not given by some authors.

We can show that

$$\frac{d^{n-m}}{dx^{n-m}}(x^2-1)^n = \frac{(n-m)!}{(n+m)!}(x^2-1)^m\frac{d^{n+m}}{dx^{n+m}}(x^2-1)^n \quad m \leq n. \ (10.1.2)$$

Writing $(x^2 - 1)^n = (x-1)^n (x+1)^n$ and using Euler's formula for the derivative of a product,

$$\frac{d^{n-m}}{dx^{n-m}}(x^2-1)^n = \sum_{p=0}^{n-m} \frac{(n-m)!}{p!(n-m-p)!} \frac{n!}{(n-p)!}(x-1)^{n-p}\frac{n!}{(m+p)!}(x+1)^{m+p}$$

$$= (n-m)!(n!)^2(x^2-1)^m \sum_{p=0}^{n-m} \frac{(x-1)^{n-m-p}(x+1)^p}{p!(n-m-p)!(n-p)!(m+p)!}$$

and

$$\frac{d^{n+m}}{dx^{n+m}}(x^2-1)^n = \sum_{p=m}^{n} \frac{(n+m)!}{p!(n+m-p)!} \frac{n!}{(n-p)!}(x-1)^{n-p}\frac{n!}{(p-m)!}(x+1)^{p-m}.$$

If we replace $p - m$ by p,

$$\frac{d^{n+m}}{dx^{n+m}}(x^2-1)^n = (n+m)!(n!)^2 \sum_{p=0}^{n-m} \frac{(x-1)^{n-m-p}(x+1)^p}{(p+m)!(n-p)!(n-m-p)!p!}.$$

We can define $P_n^{-m}(x)$ for positive m as

$$P_n^{-m}(x) = \frac{(-1)^m}{2^n n!}(1-x^2)^{-m/2}\frac{d^{n-m}}{dx^{n-m}}(x^2-1)^n = (-1)^m \frac{(n-m)!}{(n+m)!}P_n^m(x).$$

$$(10.1.3)$$

The first few associated Legendre functions are:

$P_1^0(x) = x.$
$P_1^1(x) = -\sqrt{1-x^2}.$
$P_2^0(x) = (3x^2 - 1)/2.$
$P_2^1(x) = -3x\sqrt{1-x^2}.$
$P_2^2(x) = 3(1-x^2).$
$P_3^0(x) = (5x^3 - 3x)/2.$
$P_3^1(x) = -3\sqrt{1-x^2}(5x^2-1)/2.$
$P_3^2(x) = 15x(1-x^2).$
$P_3^3(x) = -15(1-x^2)^{3/2}.$
$P_4^0(x) = (35x^4 - 30x^2 + 3)/8.$
$P_4^1(x) = -5\sqrt{1-x^2}(7x^3 - 3x)/2.$
$P_4^2(x) = 15(1-x^2)(7x^2-1)/2.$
$P_4^3(x) = -105x(1-x^2)^{3/2}.$
$P_4^4(x) = 105(1-x^2)^2.$

Graphs of $P_n^2(x)$ for $n = 2$ to $n = 5$.

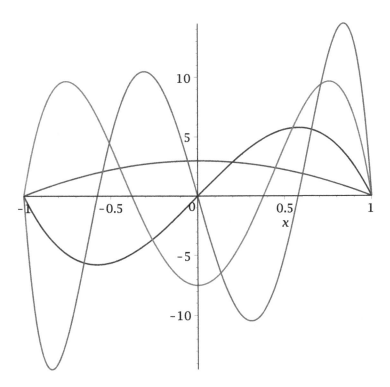

10.2 Orthogonality Relations

The associated Legendre functions satisfy the orthogonality relations

$$\int_{-1}^{1} P_n^m(x)P_{n'}^m(x)dx = \frac{2(n+m)!}{(2n+1)(n-m)!}\delta_{n,n'} \qquad m \le n, \qquad (10.2.1)$$

where $\delta_{n,n'}$ is the Kronecker delta function which equals 1 if $n = n'$ and zero otherwise. To prove this, consider with $p \ge 0$

$$\frac{(n-m)!}{(n+m)!}\int_{-1}^{1} P_n^m(x)P_{n+p}^m(x)dx$$

$$= \frac{(-1)^m}{2^{2n+p}n!\,(n+p)!}\int_{-1}^{1}\frac{d^{n-m}}{dx^{n-m}}(x^2-1)^n\frac{d^{n+m+p}}{dx^{m+n+p}}(x^2-1)^n dx.$$

We integrate by parts $n - m - p$ times, integrating the first derivative and differentiating the second. The integrated part vanishes each time owing to the factor $(x^2 - 1)$ arising from the first derivative in the integral. We obtain

$$\frac{(-1)^{n+p}}{2^{2n+p}n!\,(n+p)!} \int_{-1}^{1} \frac{d^p}{dx^p}(x^2 - 1)^n \frac{d^{2n}}{dx^{2n}}(x^2 - 1)^n dx$$

$$= \frac{(2n)!(-1)^{n+p}}{2^{2n+p}n!\,(n+p)!} \int_{-1}^{1} \frac{d^p}{dx^p}(x^2 - 1)^n dx.$$

If $p > 0$ the integral vanishes at both end points. For $p = 0$, we have

$$\frac{(2n)!}{(2^n n!)^2} \int_{-1}^{1}(1 - x^2)^n dx = \frac{(2n)!}{(2^n n!)^2} \int_{0}^{1}(1 - t)^n t^{-1/2} dt$$

$$= \frac{(2n)!}{(2^n n!)^2} B(n+1, 1/2) = \frac{(2n)!}{(2^n n!)^2} \frac{n!\Gamma(1/2)}{\Gamma(n+3/2)} = \frac{2}{2n+1},$$

where $B(p,q)$ is the Beta function $B(p,q)$. See the General Appendix for a definition of the Beta function, the Gamma function and some of their properties including the formula for $(2n)!$.

The orthogonality relation

$$\int_{-1}^{1} P_n^m(x) P_{n'}^{-m}(x) dx = (-1)^m \frac{2}{2n+1} \delta_{n,n'} \qquad (10.2.2)$$

follows from (10.2.2) and (10.1.3).

Two more orthogonality relations are:

$$\int_{-1}^{1} \frac{P_n^l(x) P_n^m(x)}{1 - x^2} dx = \delta_{l,m} \frac{(n+m)!}{(n-m)!\,m} \qquad (10.2.3)$$

and

$$\int_{-1}^{1} \frac{P_n^l(x) P_n^{-m}(x)}{1 - x^2} dx = \delta_{l,m} \frac{(-1)^m}{m}. \qquad (10.2.4)$$

The second of these follows from the first and Eq. (10.1.3).

Consider firstly the case when $m \neq l$. If $l - m$ is odd, the integrand is odd and therefore the integral is zero. Suppose now that $l = m + 2p$. Then

$$\int_{-1}^{1} \frac{P_n^l(x) P_n^m(x)}{1 - x^2} dx$$

$$= (-1)^m \int_{-1}^{1} \left[(1 - x^2)^m \frac{d^m}{dx^m} P_n(x)\right]\left[(1 - x^2)^{p-1} \frac{d^{m+2p}}{dx^{m+2p}} P_n(x)\right] dx.$$

The second factor in square brackets is a polynomial of order $n - m - 2$ and is therefore orthogonal to $(1-x^2)^m d^m P_n(x)/dx^m$. The integral is therefore zero if $l \neq m$. The evaluation when $l = m$ is given in the appendix to this chapter.

10.3 Differential Equation

The ordinary differential equation satisfied by the Associated Legendre Functions is

$$(1-x^2)\frac{d^2y}{dx^2} - 2x\frac{dy}{dx} + \left(\lambda - \frac{m^2}{1-x^2}\right)y = 0. \tag{10.3.1}$$

The point $x = 0$ is an ordinary point. This means that we can express the solution in the form of a power series $y = \sum_{n=0}^{\infty} a_n x^n$. The singularities at $x = \pm 1$ mean that the radius of convergence of this series solution is 1. Substitution of this expansion into the differential equation leads to a recurrence relation involving a_{n+2}, a_n and a_{n-2}. A simpler equation arises if we write $y = (1-x^2)^{m/2}w$. This removes the term $m^2/(1-x^2)$ from the differential equation. Then

$$(1-x^2)\frac{d^2w}{dx^2} - 2(m+1)x\frac{dw}{dx} + (\lambda - m - m^2)w = 0. \tag{10.3.2}$$

This is essentially the same equation as that satisfied by the Gegenbauer polynomials, Eq. (9.2.1). If we now substitute the expansion $w = \sum_{n=0}^{\infty} a_n x^n$ into the left-hand side of Eq. (10.3.2), we get

$$(1-x^2)\sum_{n=0}^{\infty} n(n-1)a_n x^{n-2} - 2(m+1)x\sum_{n=0}^{\infty} na_n x^n + (\lambda - m - m^2)\sum_{n=0}^{\infty} a_n x^n.$$

Equating the sum of the coefficients of x^n to zero

$$(n+2)(n+1)a_{n+2} - n(n-1)a_n - 2(m+1)na_n + (\lambda - m - m^2)a_n = 0.$$

This leads to the recurrence relation

$$\frac{a_{n+2}}{a_n} = \frac{(n+m)(n+m+1) - \lambda}{(n+2)(n+1)}. \tag{10.3.3}$$

We see that there are two power series solutions, one consisting of even powers and the other only odd powers.

The ratio test shows that both series converge for $|x| < 1$, but the test fails for $|x| = 1$. We use Gauss's test to determine the convergence in this case. (Ferrar or Phillips) For the even power series if we let $u_p = a_{2p}$, then

$$\frac{u_p}{u_{p+1}} = \frac{(2p+2)(2p+1)}{(2p+m)(2p+m+1) - \lambda} = 1 + \frac{1-m}{p} + O\left(\frac{1}{p^2}\right). \tag{10.3.4}$$

For all non-negative values of m, the coefficient of $1/p$ is less than or equal to 1 and so the series will diverge for $x = 1$. The same argument can be used to show that the odd power series diverges for $x = 1$.

If $\lambda = (N + m)(N + m + 1)$ where N is an integer, one of the series solutions will terminate and be a polynomial of order N. Let us denote this polynomial solution by $R_N^m(x)$. This polynomial satisfies the differential equation

$$(1-x^2)\frac{d^2 R_N^m(x)}{dx^2} - (2m+1)x\frac{dR_N^m(x)}{dx} + N(N+2m+1)R_N^m(x) = 0. \quad (10.3.5)$$

This equation is satisfied by the Gegenbauer polynomial $C_N^{(m+1/2)}(x)$. This means that $R_N^m(x)$ must be a multiple of this polynomial. The actual multiple is given in Eq. (10.1.1).

10.4   Orthogonality

The differential equation (10.3.2) for $R_N^m(x)$ with $\lambda = (N + m)(N + m + 1)$ can be written in Sturm Liouville form

$$\frac{d}{dx}\left((1 - x^2)^{m+1}\frac{dR_N^m(x)}{dx}\right) = -(N + m)(N + m + 1)(1 - x^2)^m R_N^m(x).$$
$$(10.4.1)$$

If we multiply this by $R_M^m(x)$ and integrate from -1 to 1, the left-hand side becomes

$$\int_{-1}^{1} R_M^m(x)\frac{d}{dx}\left((1 - x^2)^{m+1}\frac{dR_N^m(x)}{dx}\right)dx.$$

On integrating this by parts, noting that the integrated term vanishes at both end points, we find that

$$\int_{-1}^{1}(1 - x^2)^{m+1}\left[\frac{dR_M^m(x)}{dx}\right]\left[\frac{dR_N^m(x)}{dx}\right]dx$$

$$= -(N + m)(N + m + 1)\int_{-1}^{1}(1 - x^2)^m R_M^m(x)R_N^m(x)dx. \quad (10.4.2)$$

If we follow the same procedure as above but with M and N interchanged we will produce an equation which is the same as that above on the left-hand side but with the coefficient $(M + m)(M + m + 1)$ on the right-hand side. If we subtract one of these equations from the other, we get

$$\left[(M+m)(M+m+1)-(N+m)(N+m+1)\right]\int_{-1}^{1}(1-x^2)^m R_M^m(x)R_N^m(x)dx = 0.$$

Thus for $N \neq M$

$$\int_{-1}^{1} (1 - x^2)^m R_M^m(x) R_N^m(x) dx = 0. \qquad (10.4.3)$$

The polynomials $R_N^m(x)$ thus satisfy the same orthogonality relations as the Gegenbauer polynomials $C_N^{(m+1/2)}(x)$ and must therefore be multiples of them.

It follows then that the functions $(1 - x^2)^{m/2} R_N^m(x)$ are multiples of $P_{N+m}^m(x)$.

10.5   Generating Function

The generating function

$$\sum_{p=0}^{\infty} t^p P_{m+p}^m(x) = \frac{(-1)^m (2m)!}{2^m m!} \frac{(1 - x^2)^{m/2}}{(1 - 2xt + t^2)^{m+1/2}} \qquad |t| < 1 \qquad (10.5.1)$$

follows from the generating function for the Gegenbauer polynomials.

10.6   Recurrence Relations

A number of recurrence relations can be derived from those for the Gegenbauer polynomials.

From Eq. (9.7.2) we deduce that

$$\sqrt{1 - x^2}(n - m + 1) P_{n+1}^m(x) = P_n^{m+1}(x) - x P_{n+1}^{m+1}(x). \qquad (10.6.1)$$

This result can be obtained by differentiating (10.5.1) with respect to t,

$$\sum_{p=0}^{\infty} (p + 1) t^p P_{m+p+1}^m(x) = \frac{(t - x)}{\sqrt{1 - x^2}} \sum_{p=0}^{\infty} t^p P_{m+p+1}^{m+1}(x)$$

and equating coefficients of t^p on both sides of the equation.

From Eq. (9.7.3) we deduce that

$$(n - m + 1) P_{n+1}^m(x) = (2n + 1) x P_n^m(x) - (n + m) P_{n-1}^m(x). \qquad (10.6.2)$$

Alternatively, differentiating (10.5.1) with respect to t, multiplying through by $(1 - 2xt + t^2)$ and substituting the left-hand side of (10.5.1) gives

$$(1 - 2xt + t^2) \sum_{p=0}^{\infty} p t^{p-1} P_{m+p}^m(x) = (m + 1/2)(2x - 2t) \sum_{p=0}^{\infty} t^p P_{m+p}^m(x).$$

Equating the coefficients of t^p on both sides of this equation leads to (10.6.2).

From Eq. (9.7.4),

$$\sqrt{1 - x^2}P_n^m(x) = \frac{1}{2n + 1}\left[P_{n-1}^{m+1}(x) - P_{n+1}^{m+1}(x)\right]. \tag{10.6.3}$$

Alternatively, we can use the generating function to show that

$$\left(t\frac{\partial}{\partial t} + m + \frac{1}{2}\right)\sum_{p=0}^{\infty} t^p P_{m+p}^m(x)$$

$$= -\frac{(-1)^m[2(m + 1)]!}{2^{m+2}(m + 1)!}(1 - x^2)^{m/2}\frac{1 - t^2}{(1 - 2xt + t^2)^{m+3/2}}$$

$$= (t^2 - 1)\sum_{p=0}^{\infty} t^p P_{m+p+1}^{m+1}(x)/(2\sqrt{1 - x^2}).$$

Equating the coefficients of t^p leads to (10.6.3).

From Eq. (9.7.5),

$$\sqrt{1 - x^2}P_n^{m+1}(x) = (n - m)xP_n^m(x) - (n + m)P_{n-1}^m(x). \tag{10.6.4}$$

From Eq. (9.7.6),

$$2mxP_n^m(x) = -\sqrt{1 - x^2}\left[P_n^{m+1}(x) + (n+m)(n-m+1)P_n^{m-1}(x)\right]. \tag{10.6.5}$$

From Eq. (9.8.7),

$$(2n + 1)\sqrt{1 - x^2}P_n^m(x)$$

$$= (n - m + 1)(n - m + 2)P_{n+1}^{m-1}(x) - (n + m - 1)(n + m)P_{n-1}^{m-1}(x). \tag{10.6.6}$$

10.7 Differential Relations

A number of expressions for the derivative of the associated Legendre function follow from those for the Gegenbauer polynomials. Firstly from Eq. (9.8.1)

$$(x^2 - 1)\frac{d}{dx}P_n^m(x) = nxP_n^m(x) - (n + m)P_{n-1}^m(x). \tag{10.7.1}$$

If we combine this with Eq. (10.6.4) we get

$$(x^2 - 1)\frac{d}{dx}P_n^m(x) = \sqrt{1 - x^2}P_n^{m+1}(x) + mxP_n^m(x). \tag{10.7.2}$$

Combining Eq. (10.7.1) with (10.6.5) gives

$$(1 - x^2)\frac{d}{dx}P_n^m(x) = (n + m)(n - m + 1)\sqrt{1 - x^2}P_n^{m-1}(x) + mxP_n^m(x). \tag{10.7.3}$$

10.8 Step Up and Step Down Operators

These operators can be easily derived from relations (10.6.2) and (10.7.1).

$$P_{n-1}^m(x) = S_n^- P_n^m(x) = \left\{ \frac{nx}{n+m} + \frac{1-x^2}{n+m} \frac{d}{dx} \right\} P_n^m(x) \qquad (10.8.1)$$

and

$$P_{n+1}^m(x) = S_n^+(x) P_n^m(x) = \left\{ \frac{(n+1)x}{n-m+1} - \frac{1-x^2}{n-m+1} \frac{d}{dx} \right\} P_n^m(x). \quad (10.8.2)$$

10.9 Appendix

Consider the integral

$$I_{m,n,p} = \int_{-1}^{1} (1-x^2)^{m-p-1} \left[\frac{d^{m-p}}{dx^{m-p}} P_n(x) \right] \left[\frac{d^m}{dx^m} P_n(x) \right] dx.$$

We note that if p is odd, the integrand is odd and therefore this integral is zero. Integrating by parts, we find

$$I_{m,n,p} = \left[(1-x^2)^{m-p-1} \left[\frac{d^{m-p-1}}{dx^{m-p-1}} P_n(x) \right] \left[\frac{d^m}{dx^m} P_n(x) \right] \right]_{-1}^{1}$$

$$- \int_{-1}^{1} dx \frac{d}{dx} \left\{ (1-x^2)^{m-p-1} \frac{d^m}{dx^m} P_n(x) \right\} \frac{d^{m-p-1}}{dx^{m-p-1}} P_n(x).$$

If m is odd and $p = m - 1$, the integrated part becomes

$$2P_n(x) \frac{d^m}{dx^m} P_n(x) \bigg|_{x\to 1} = \frac{P_n(x)}{2^{n-1} n!} \frac{d^{m+n}}{dx^{m+n}} (x^2-1)^n \bigg|_{x\to 1} = \frac{(n+m)!}{2^{m-1}(n-m)! \, m!},$$

and the integral is zero, the polynomial $d^{m+1} P_n(x)/dx^{m+1}$ of order $n-m-1$ being orthogonal to the Legendre polynomial $P_n(x)$.

If $p < m - 1$, the integrated term is zero and the integral becomes

$$- \int_{-1}^{1} dx (1-x^2)^{m-p-1} \left[\frac{d^{m-p-1}}{dx^{m-p-1}} P_n(x) \right] \left[\frac{d^{m+1}}{dx^{m+1}} P_n(x) \right]$$

$$+2(m-p-1) \int_{-1}^{1} dx (1-x^2)^{m-p-2} x \left[\frac{d^{m-p-1}}{dx^{m-p-1}} P_n(x) \right] \left[\frac{d^m}{dx^m} P_n(x) \right].$$

The first of these integrals is zero since $d^{m-p-1} P_n(x)/dx^{m-p-1}$ is a multiple of the Gegenbauer polynomial $C_{n-m+p-1}^{m-p-1/2}(x)$ which with the weight factor

$(1 - x^2)^{m-p-1}$ is orthogonal to the $(n - m - 1)$th order polynomial $d^{m+1}P_n(x)/dx^{m+1}$.

We now integrate the second integral by parts to obtain

$$2(m - p - 1)\left[(1 - x^2)^{m-p-2}x\left[\frac{d^{m-p-2}}{dx^{m-p-2}}P_n(x)\right]\left[\frac{d^m}{dx^m}P_n(x)\right]\right]_{-1}^1$$

$$-2(m - p - 1)\int_{-1}^1 dx\,\frac{d^{m-p-2}}{dx^{m-p-2}}P_n(x)\frac{d}{dx}\left\{(1 - x^2)^{m-p-2}x\frac{d^m}{dx^m}P_n(x)\right\}.$$

If m is even and $p = m - 2$, the integrated part is

$$4xP_n(x)\frac{d^m}{dx^m}P_n(x)\bigg|_{x\to1} = \frac{xP_n(x)}{2^{n-2}n!}\frac{d^{m+n}}{dx^{m+n}}(x^2 - 1)^n\bigg|_{x\to1} = \frac{(n + m)!}{2^{m-2}(n - m)!m!},$$

and the integral is zero.

If $p < m - 2$, the integrated term is zero and the integral becomes

$$-\int_{-1}^1 dx(1 - x^2)^{m-p-2}\frac{d^{m-p-2}}{dx^{m-p-2}}P_n(x)\frac{d}{dx}\left\{x\frac{d^m}{dx^m}P_n(x)\right\}$$

$$+2(m - p - 2)\int_{-1}^1 dx\,x^2(1 - x^2)^{m-p-3}\left[\frac{d^{m-p-2}}{dx^{m-p-2}}P_n(x)\right]\left[\frac{d^m}{dx^m}P_n(x)\right].$$

The first integral is zero since $d^{m-p-2}P_n(x)/dx^{m-p-2}$ is proportional to the Gegenbauer polynomial $C_{n-m+p+2}^{m-p-3/2}(x)$, which with the weight factor $(1 - x^2)^{m-p-2}$ is orthogonal to the other factor in the integral which is an $(n - m)$th order polynomial.

In the second integral write $x^2 = 1 - (1 - x^2)$. The integral with $(1 - x^2)$ is zero. This leads to the recurrence relation

$$I_{m,n,p} = 2(m - p - 1)2(m - p - 2)I_{m,n,p+2}.$$

Application of this recurrence relation to $I_{m,n,0}$ and the using the results when $p = m-1$ if m is odd or $p = m-2$ when m is even leads to Eq. (10.1.6).

References

Copson E T, Theory of Functions of a Complex Variable, Oxford University Press, 1955.

Courant R and Hilbert D, Methods of Mathematical Physics Vol 1, Interscience Publishers, 1953.

Dunster T M Digital Chapter 14, NIST Handbook of Mathematical Functions, eds Olver W J, Lozier D W, Boisvert R F and Clark C W, NIST and Cambridge University Press, 2009.

Dennery P and Krzywicki A, Mathematics for Physicists, Harper and Rowe, 1967.

Erdelyi A, Higher Transcendental Functions Vol 2, McGraw-Hill, 1953.

Ferrar W L, Textbook of Convergence, Oxford University Press, 1938.

Hochstrasser Urs W, Orthogonal Polynomials, Chapter 22, Handbook of Mathematical Functions, Eds. Abramowitz M and Stegun I A, Dover, 1970.

Koornwinder T H, Wong R, Koekoek R and Swarttouw R F, Chapter 18, NIST Handbook of Mathematical Functions, Eds. Olver W J, Lozier D W, Boisvert R F and Clark C W, NIST and Cambridge University Press, 2009.

Macrobert T M, Functions of a Complex Variable, Macmillan, 1933.

Mandl F, Quantum Mechanics, Butterworth Scientific Publications, 1957.

Phillips E G, A Course of Analysis, Cambridge University Press, 1930.

Pauling L and Wilson E B, Introduction to Quantum Mechanics, McGraw-Hill, 1935.

Schiff L I, Quantum Mechanics, McGraw-Hill, 1955.

Sneddon I N, Special Functions of Mathematical Physics and Chemistry, Oliver and Boyd, 1956.

Stegun I A, Chapter 8, Handbook of Mathematical Functions, Eds. Abramowitz M and Stegun I A, Dover, 1970.

Whittaker E T and Watson G N, A Course of Modern Analysis, Cambridge University Press, 1963.

Chapter 11

Jacobi Polynomials

11.1 Introduction

The Jacobi polynomials, $P_n^{(\alpha,\beta)}(x)$, are the most general of the classical orthogonal polynomials in the domain $-1 \leq x \leq 1$. All of the other clasical orthogonal polynomials in this domain are special cases. Important amongst these are:

$\alpha = \beta = 0$, Legendre polynomials $P_n(x) = P_n^{(0,0)}(x)$.

$\alpha = \beta$, Gegenbauer or Ultraspherical polynomials:

$$C_n^{(\alpha)}(x) = \frac{\Gamma(2\alpha + n)\Gamma(\alpha + 1/2)}{\Gamma(2\alpha)\Gamma(\alpha + n + 1/2)} P_n^{(\alpha-1/2,\alpha-1/2)}(x). \qquad (11.1.1)$$

The Chebyshev polynomials of the first and second kinds are each special cases of the Gegenbauer polynomials as are the Legendre polynomials. The Chebyshev polynomials of the third and fourth kinds are related by

$$V_n(x) = \frac{n!\Gamma(1/2)}{\Gamma(n + 1/2)} P_n^{(-1/2,1/2)}(x). \qquad (11.1.2)$$

$$W_n(x) = \frac{n!\Gamma(1/2)}{\Gamma(n + 1/2)} P_n^{(1/2,-1/2)}(x). \qquad (11.1.3)$$

Jacobi polynomials can be obtained using the Gram-Schmidt orthogonalisation process for polynomials in the domain $(-1, 1)$ with weight function $(1-x)^\alpha(1+x)^\beta$, where $\alpha, \beta > -1$. To complete the definition, $P_n^{(\alpha,\beta)}(1)$ is set equal to $\Gamma(n + \alpha + 1)/[\Gamma(\alpha + 1)n!]$.

The first few Jacobi polynomials are:

$P_0^{(\alpha,\beta)}(x) = 1.$

$P_1^{(\alpha,\beta)}(x) = [\alpha - \beta + (\alpha + \beta + 2)x]/2.$

$P_2^{(\alpha,\beta)}(x) = \Big\{ -4 - (\alpha + \beta) + (\alpha - \beta)^2 + 2(\alpha - \beta)(\alpha + \beta + 3)x$

$\qquad\qquad + (\alpha + \beta + 4)(\alpha + \beta + 3)x^2 \Big\}/8.$

11.2　Differential Equation

The ordinary differential equation satisfied by Jacobi polynomials is

$$(1 - x^2)\frac{d^2y}{dx^2} + [\beta - \alpha - (\alpha + \beta + 2)x]\frac{dy}{dx} + \lambda y = 0. \qquad (11.2.1)$$

The point $x = 0$ is an ordinary point. This means that we can express the solution in the form of a power series $y = \sum_{n=0}^{\infty} a_n x^n$.

Substituting into the differential equation, we get

$$(1 - x^2)\sum_{n=0}^{\infty} n(n-1)a_n x^{n-2} + [\beta - \alpha - (\alpha + \beta + 2)x]\sum_{n=0}^{\infty} a_n n x^{n-1}$$

$$+\lambda\sum_{n=0}^{\infty} a_n x^n = 0.$$

If we equate the coefficient of x^n to zero in this equation, we get

$$(n+2)(n+1)a_{n+2} + (\beta - \alpha)(n+1)a_{n+1} + [\lambda - (\alpha + \beta + 2)n - n(n-1)]a_n = 0.$$

This leads to the recurrence relation

$$a_{n+2} = \frac{\alpha - \beta}{n+2}a_{n+1} + \frac{n(n+1) + (\alpha + \beta)n - \lambda}{(n+2)(n+1)}a_n. \qquad (11.2.2)$$

This gives us two linearly independent solutions, each solution, in general containing both even and odd powers of x. If $\alpha = \beta$, we have solutions with either even or odd powers of x.

If $\lambda = m(m + 1 + \alpha + \beta)$, there will be one solution which is an mth order polynomial. For this solution

$$a_{m-1} = \frac{m(\alpha - \beta)}{(\alpha + \beta + 2m)}a_m. \qquad (11.2.3)$$

The other linearly independent solution will be an infinite series in powers of x.

We are now going to show that the polynomial solutions $R_n(x)$ satisfy the orthogonality relations for the Jacobi polymonials and must therefore be multiples of them.

11.3　Orthogonality

The differential equation (11.2.1) for $R_m(x)$ can be written in Sturm Liouville form.

$$\frac{d}{dx}\left[(1-x)^{\alpha+1}(1+x)^{\beta+1}\frac{dR_m(x)}{dx}\right] = -m(m+1+\alpha+\beta)(1-x)^{\alpha}(1+x)^{\beta}R_m(x). \qquad (11.3.1)$$

If we multiply this by $R_n(x)$ and integrate from -1 to 1, the left-hand side becomes:

$$\int_{-1}^{1} R_n(x) \frac{d}{dx} \left[(1-x)^{\alpha+1} (1+x)^{\beta+1} \frac{dR_m(x)}{dx} \right] dx.$$

On integrating by parts and noting that since $\alpha > -1$ and $\beta > -1$, the integrated term vanishes at both end points, we find that

$$\int_{-1}^{1} (1-x)^{\alpha+1} (1+x)^{\beta+1} R_n'(x) R_m'(x) dx$$

$$= m(m+1+\alpha+\beta) \int_{-1}^{1} (1-x)^{\alpha} (1+x)^{\beta} R_n(x) R_m(x) dx.$$

If we follow the same procedure as above but with m and n interchanged we will produce an equation which is the same as that above on the left-hand side and the same integral on the right but with a coefficient $n(n+1+\alpha+\beta)$ instead of $m(m+1+\alpha+\beta)$. If we subtract one of these equations from the other we get

$$(m-n)(m+n+1+\alpha+\beta) \int_{-1}^{1} (1-x)^{\alpha} (1+x)^{\beta} R_n(x) R_m(x) dx = 0.$$

Thus for $n \neq m$

$$\int_{-1}^{1} (1-x)^{\alpha} (1+x)^{\beta} R_n(x) R_m(x) dx = 0. \qquad (11.3.2)$$

The polynomials $R_m(x)$ thus satisfy the same orthogonality relations as the Jacobi polynomials and therefore must be multiples of them. If we multiply $R_n(x)$ by a constant so that $R_n(1) = \Gamma(\alpha+n+1)/[\Gamma(\alpha+1)n!]$ we obtain the Jacobi polynomial $P_n^{(\alpha,\beta)}(x)$.

Since x^n can be represented as a linear combination of $P_p^{(\alpha,\beta)}(x)$ for $0 \leq p \leq n$, we can deduce that

$$\int_{-1}^{1} (1-x)^{\alpha} (1+x)^{\beta} x^n P_m^{(\alpha,\beta)}(x) dx = 0 \qquad \text{for all } n < m. \qquad (11.3.3)$$

11.4 Derivative Property

The derivative of $P_N^{(\alpha,\beta)}(x)$, $P_N^{(\alpha,\beta)}(x)'$ is a polynomial of order $N-1$. We now show that $P_N^{(\alpha,\beta)}(x)'$ is a multiple of $P_{N-1}^{(\alpha+1,\beta+1)}(x)$. If we differentiate the differential equation for $P_N^{(\alpha,\beta)}(x)$, Eq. (11.2.1), we find

$$(1-x^2) \frac{d^2 y'}{dx^2} + \left[\beta - \alpha - (\alpha+\beta+4)x \right] \frac{dy'}{dx} + (N-1)(N+\alpha+\beta+2)y' = 0.$$

This equation is satisfied by $P_{N-1}^{(\alpha+1,\beta+1)}(x)$. This means that $P_N^{(\alpha,\beta)}(x)' = CP_{N-1}^{(\alpha+1,\beta+1)}(x)$. We can find the value of C by equating the coefficients of x^{N-1} on both sides. We find

$$C = \frac{\alpha + \beta + N + 1}{2}.$$

We have used the value of k_N, the coefficient of x^N in $P_N^{(\alpha,\beta)}(x)$ derived in section 11.6 below.

We now show that the pth derivative of $P_N^{(\alpha,\beta)}(x)$ is a polynomial of order $N - p$ which is a multiple of $P_{N-p}^{(\alpha+p,\beta+p)}(x)$. If we differentiate Eq. (11.2.1) with $\lambda = N(N + \alpha + \beta + 1)$ p times and denote $d^p y/dx^p$ by ϕ, we find

$$(1 - x^2)\frac{d^2\phi}{dx^2} + \left[\alpha - \beta - (\alpha + \beta + 2p + 2)x\right]\frac{d\phi}{dx}$$

$$+(N - p)(N + p + \alpha + \beta + 1)\phi = 0.$$

This equation is satisfied by $\phi = P_{N-p}^{(\alpha+p,\beta+p)}(x)$. As before, we can find the constant by looking at the coefficients of x^{N-p}:

$$\frac{d^p}{dx^p} P_N^{(\alpha,\beta)}(x) = \frac{\Gamma(\alpha + \beta + N + p + 1)}{2^p\,\Gamma(\alpha + \beta + N + 1)} P_{N-p}^{(\alpha+p,\beta+p)}(x).$$

11.5 Rodrigues Formula

If we differentiate the product $(1 - x)^{n+\alpha}(1 + x)^{n+\beta}$ m times, for $m < n$, we see that with $\alpha > -1$ and $\beta > -1$,

$$\frac{d^m}{dx^m}\left[(1 - x)^{n+\alpha}(1 + x)^{n+\beta}\right] = 0 \qquad \text{when } x = \pm 1.$$

From this result, we see that

$$\int_{-1}^{1} \frac{d^n}{dx^n}\left[(1 - x)^{n+\alpha}(1 + x)^{n+\beta}\right] dx = 0.$$

If we integrate by parts we see that

$$\int_{-1}^{1} x\frac{d^n}{dx^n}\left[(1 - x)^{n+\alpha}(1 + x)^{n+\beta}\right] dx = 0 \quad \text{provided that } n > 1,$$

and integrating by parts m times we see that

$$\int_{-1}^{1} x^m\frac{d^n}{dx^n}\left[(1-x)^{n+\alpha}(1+x)^{n+\beta}\right] dx = 0 \quad \text{provided that } n > m. \quad (11.5.1)$$

Let us define the nth order polynomial
$$Q_n(x) = \frac{1}{(1-x)^\alpha(1+x)^\beta}\frac{d^n}{dx^n}\left[(1-x)^{\alpha+n}(1+x)^{\beta+n}\right],$$
then
$$\int_{-1}^1 (1-x)^\alpha(1+x)^\beta x^m Q_n(x)\,dx = 0 \qquad m < n$$
is orthogonal to x^m for all values of $m < n$. This means that $Q_m(x)$ is orthogonal to $Q_n(x)$ for all values of $m < n$. These polynomials must therefore be multiples of the Jacobi polynomials we found earlier. We can find out what this multiple is by evaluating $Q_n(1)$.
$$Q_n(1) = \lim_{x\to 1}\left\{\frac{1}{(1-x)^\alpha(1+x)^\beta}\frac{d^n}{dx^n}\left((1-x)^{n+\alpha}(1+x)^{n+\beta}\right)\right\}.$$
The only term which contributes in the limit $x \to 1$ is the nth derivative of $(1-x)^{n+\alpha}$ multiplied by $(1+x)^{n+\beta}$. Thus
$$Q_n(1) = (n+\alpha)(n+\alpha-1)...(\alpha+1)2^n(-1)^n.$$
Therefore, since $P_n^{(\alpha,\beta)}(1) = \Gamma(\alpha+n+1)/[n!\,\Gamma(\alpha+1)]$,
$$P_n^{(\alpha,\beta)}(x) = \frac{-1^n}{2^n n!\,(1-x)^\alpha(1+x)^\beta}\frac{d^n}{dx^n}\left[(1-x)^{\alpha+n}(1+x)^{\beta+n}\right]. \quad (11.5.2)$$
We can now evaluate
$$h_N = \int_{-1}^1 (1-x)^\alpha(1+x)^\beta[P_N^{[\alpha,\beta)}(x)]^2 dx.$$
Using the formula (11.5.2),
$$h_N = \frac{(-1)^N}{2^N N!}\int_{-1}^1 P_N^{(\alpha,\beta)}(x)\frac{d^N}{dx^N}\left[(1-x)^{\alpha+N}(1+x)^{\beta+N}\right]dx.$$
Integrating by parts N times noting that
$d^m/dx^m\left[(1-x)^{\alpha+N}(1+x)^{\beta+N}\right] = 0$ for $x = \pm 1$ for $m < N$, so that
$$h_N = \frac{1}{2^N N!}\int_{-1}^1 (1-x)^{\alpha+N}(1+x)^{\beta+N}\frac{d^N}{dx^N}P_N^{(\alpha,\beta)}(x)dx$$
$$= \frac{1}{2^{2N} N!}\frac{\Gamma(\alpha+\beta+2N+1)}{\Gamma(\alpha+\beta+N+1)}\int_{-1}^1 (1-x)^{\alpha+N}(1+x)^{\beta+N}dx,$$
on substituting the value of k_N from (11.6.4) below. Writing $x = 1 - 2y$ gives
$$h_N = \frac{2^{\alpha+\beta+1}}{N!}\frac{\Gamma(\alpha+\beta+2N+1)}{\Gamma(\alpha+\beta+N+1)}\int_0^1 y^{\alpha+N}(1-y)^{\beta+N}dy$$
$$= \frac{2^{\alpha+\beta+1}}{N!}\frac{\Gamma(\alpha+\beta+2N+1)}{\Gamma(\alpha+\beta+N+1)}B(\alpha+N+1,\beta+N+1)$$
$$= \frac{2^{\alpha+\beta+1}}{N!}\frac{\Gamma(\alpha+\beta+2N+1)}{\Gamma(\alpha+\beta+N+1)}\frac{\Gamma(\alpha+N+1)\Gamma(\beta+N+1)}{\Gamma(\alpha+\beta+2N+2)}$$
$$= \frac{2^{\alpha+\beta+1}}{N!}\frac{\Gamma(\alpha+N+1)\Gamma(\beta+N+1)}{\Gamma(\alpha+\beta+N+1)(\alpha+\beta+2N+1)}.$$

11.6   Explicit Expression

If we use the formula for the nth derivative of the product in Eq. (11.5.2),

$$P_n^{(\alpha,\beta)}(x) = \frac{1}{2^n} \sum_{m=0}^{n} \frac{\Gamma(n+\alpha+1)(x-1)^m}{\Gamma(m+\alpha+1)(n-m)!} \frac{\Gamma(n+\beta+1)(x+1)^{n-m}}{\Gamma(n-m+\beta+1)m!}.$$

(11.6.1)

From this expression we can see that $P_n^{(\alpha,\beta)}(-x) = (-1)^n P_n^{(\beta,\alpha)}(x)$.

We can obtain an expansion in powers of $y = (x-1)$ by writing $x+1 = y+2$ and expanding powers of $(x+1)$. We find

$$P_n^{(\alpha,\beta)}(x) = \frac{1}{2^n} \sum_{m=0}^{n} \frac{\Gamma(n+\alpha+1)\,y^m}{\Gamma(\alpha+1+m)(n-m)!} \frac{\Gamma(n+\beta+1)}{\Gamma(n-m+\beta+1)m!}$$

$$\times \sum_{p=0}^{n-m} \frac{(n-m)!}{(n-m-p)!\,p!} 2^{n-m-p} y^p.$$

We now change the summations by writing $q = m+p$ and summing over m before summing over q. Now m goes from 0 to q and then q goes from 0 to n.

$$P_n^{(\alpha,\beta)}(x) = \sum_{q=0}^{n} \frac{(y/2)^q}{(n-q)!} \sum_{m=0}^{q} \frac{\Gamma(n+\alpha+1)\Gamma(n+\beta+1)}{\Gamma(\alpha+1+m)\Gamma(n-m+\beta+1)m!(q-m)!}.$$

The sum over m can be written

$$\frac{\Gamma(n+\alpha+1)}{\Gamma(\alpha+1)q!} \left[1 + \frac{(n+\beta)q}{(\alpha+1).1} + \frac{(n+\beta)(n+\beta-1)q(q-1)}{(\alpha+1)(\alpha+2).2!} + \cdots \right]$$

$$= \frac{\Gamma(n+\alpha+1)}{\Gamma(\alpha+1)q!} \, {}_2F_1(-n-\beta,-q;\alpha+1;1)$$

$$= \frac{\Gamma(n+\alpha+1)\Gamma(\alpha+\beta+n+1+q)}{\Gamma(\alpha+1+q)\Gamma(\alpha+\beta+n+1)q!}.$$

We have used Vandermonde's theorem to evaluate the hypergeometric polynomial

$$_2F_1(a,-n;c;1) = \frac{(c-a)_n}{c_n} = \frac{\Gamma(c-a+n)\Gamma(c)}{\Gamma(c-a)\Gamma(c+n)}.$$

This relation is proved in the General Appendix.

Substituting into the expression for $P_n^{(\alpha,\beta)}(x)$ gives

$$P_n^{(\alpha,\beta)}(x) = \frac{\Gamma(n+\alpha+1)}{\Gamma(\alpha+1)n!}\left[1 + \frac{(\alpha+\beta+n+1)n}{(\alpha+1).1}\left(\frac{y}{2}\right)\right.$$

$$\left. + \frac{(\alpha+\beta+n+1)(\alpha+\beta+n+2)n(n-1)}{(\alpha+1)(\alpha+2).2!}\left(\frac{y}{2}\right)^2 + ... \right]$$

$$= \frac{\Gamma(n+\alpha+1)}{\Gamma(\alpha+1)n!}\,_2F_1(-n,\alpha+\beta+n+1;\alpha+1;-y/2) \qquad (11.6.2)$$

$$= \frac{(-1)^n}{n!}\frac{d^n}{dt^n}\left(\frac{(1-t)^{\alpha+n}}{[1-(1-x)t/2]^{\alpha+\beta+n+1}}\right)\bigg|_{t\to 0}, \qquad (11.6.3)$$

on using Eq. (H8) from the General Appendix.

The coefficient of x^n, k_n, can be calculated directly from Eq. (11.6.2) or Eq. (11.6.3) above

$$k_n = \frac{\Gamma(n+\alpha+1)}{\Gamma(\alpha+1)n!}\frac{n!\,\Gamma(\alpha+\beta+2n+1)\Gamma(\alpha+1)}{\Gamma(\alpha+\beta+n+1)\Gamma(n+\alpha+1)n!\,2^n}.$$

Alternatively, from (11.6.1),

$$k_n = \frac{1}{2^n}\sum_{m=0}^{n}\frac{\Gamma(n+\alpha+1)}{\Gamma(n+\alpha+1-m)m!}\frac{\Gamma(n+\beta+1)}{\Gamma(m+\beta+1)(n-m)!}$$

$$= \frac{1}{2^n n!}\frac{\Gamma(n+\beta+1)}{\Gamma(\beta+1)}\left[1 + \frac{(n+\alpha)n}{(\beta+1).1} + \frac{(n+\alpha)(n+\alpha-1)n(n-1)}{(\beta+1)(\beta+2).1.2} + ... \right]$$

$$= \frac{1}{2^n n!}\frac{\Gamma(n+\beta+1)}{\Gamma(\beta+1)}\,_2F_1(-n-\alpha,-n;\beta+1;1)$$

$$= \frac{1}{2^n n!}\frac{\Gamma(n+\beta+1)}{\Gamma(\beta+1)}\frac{\Gamma(\beta+1)\Gamma(\alpha+\beta+2n+1)}{\Gamma(\beta+n+1)\Gamma(\alpha+\beta+n+1)},$$

$$k_n = \frac{\Gamma(\alpha+\beta+2n+1)}{2^n\,n!\,\Gamma(\alpha+\beta+n+1)}, \qquad (11.6.4)$$

where we have again used Vandermonde's theorem to evaluate the hypergeometric function $_2F_1(-n-\alpha,-n;\beta+1;1)$.

The coefficient of x^{n-1}, k_n', can be obtained by multiplying (11.6.4) by the recurrence relation for the series solution of the differential equation (11.3.3). We find

$$k_n' = \frac{(\alpha-\beta)\Gamma(\alpha+\beta+2n)}{2^n\,(n-1)!\,\Gamma(\alpha+\beta+n+1)}. \qquad (11.6.5)$$

11.7 Generating Function

Firstly let $R = \sqrt{1 - 2xt + t^2}$ and then define $w(x,t)$ by

$$w(x,t) = \frac{1}{R(1 - t + R)^\alpha (1 + t + R)^\beta}. \tag{11.7.1}$$

We can show that

$$(1-x^2)\frac{\partial^2 w}{\partial x^2} + [\beta - \alpha - (\alpha + \beta + 2)x]\frac{\partial w}{\partial x} + t\frac{\partial}{\partial t}\left(t^{-\alpha-\beta}\frac{\partial(t^{\alpha+\beta+1}w)}{\partial t}\right). \tag{11.7.2}$$

This is most easily done with the help of a computer algebra package such as Maple.

Let us define $\phi_n(x)$ by

$$\sum_{n=0}^{\infty} \phi_n(x)t^n = w(x,t) = \frac{1}{R(1 - t + R)^\alpha(1 + t + R)^\beta} \qquad |t| < 1. \tag{11.7.3}$$

On expanding the right-hand side and collecting all the terms in t^n, we see that $\phi_n(x)$ are nth order polynomials in x.

If we substitute the left-hand side of (11.7.3) into (11.7.2),

$$\sum_{n=0}^{\infty} t^n\left\{(1 - x^2)\frac{d^2\phi_n(x)}{dx} + (\beta - \alpha - (\alpha + \beta + 2)x)\frac{d\phi_n(x)}{dx}\right\}$$

$$= -\sum_{n=0}^{\infty} \phi_n(x)t\frac{d}{dt}\left(t^{-\alpha-\beta}\frac{dt^{\alpha+\beta+n+1}}{dt}\right) = -\sum_{n=0}^{\infty} n(n + \alpha + \beta + 1)t^n\phi_n(x). \tag{11.7.4}$$

If we now equate the coefficients of t^n on both sides of this equation,

$$(1-x^2)\frac{d^2\phi_n(x)}{dx^2} + [\beta - \alpha - (\alpha + \beta + 2)x]\frac{d\phi_n(x)}{dx} = -n(n + \alpha + \beta + 1)\phi_n(x),$$

we see that $\phi_n(x)$ satisfy the differential equation for Jacobi polynomials and must therefore be multiples of them. We can find out what this multiple is by putting $x = 1$ in (11.7.3). We find:

$$2^{\alpha+\beta}w(1,t) = \frac{1}{(1-t)^{\alpha+1}} = \sum_{n=0}^{\infty} \frac{\Gamma(\alpha + n + 1)}{\Gamma(\alpha + 1)n!}t^n = \sum_{n=0}^{\infty} P_n^{(\alpha,\beta)}(1)t^n.$$

It follows that

$$w(x,t) = \frac{1}{R(1 - t + R)^\alpha(1 + t + R)^\beta} = \frac{1}{2^{\alpha+\beta}}\sum_{n=0}^{\infty} P_n^{(\alpha,\beta)}(x)t^n. \tag{11.7.5}$$

11.8 Recurrence Relations

In the introduction chapter it was shown that for any set of orthogonal polynomials there is a recurrence relation of the form:

$$R_{n+1}(x) - axR_n(x) = bR_n(x) + cR_{n-1}(x). \tag{11.8.1}$$

The coefficient a is chosen so that the term $k_{n+1}x^{n+1}$ in $R_{n+1}(x)$ cancels the term $k_n x^{n+1}$ in $xR_n(x)$. Thus for the Jacobi polynomials

$$a = \frac{k_{n+1}}{k_n} = \frac{(\alpha + \beta + 2n + 2)(\alpha + \beta + 2n + 1)}{2(n+1)(\alpha + \beta + n + 1)}. \tag{11.8.2}$$

The coefficients b and c can be determined by equating the coefficients of x^n and x^{n-1} in Eq. (11.8.1), or by writing the recurrence relation Eq. (11.8.1) for $x = 1$ and $x = -1$ Putting $x = 1$ and cancelling through by $P_{n-1}^{(\alpha,\beta)}(1) = \Gamma(n + \alpha)/[(n-1)! \Gamma(\alpha + 1)]$,

$$\frac{(\alpha + n + 1)(\alpha + n)}{n(n+1)} - a\frac{\alpha + n}{n} = b\frac{\alpha + n}{n} + c.$$

For $x = -1$, after cancelling through by $P_{n-1}^{(\alpha,\beta)}(-1) = (-1)^{n-1}\Gamma(n + \beta)/[(n-1)! \Gamma(\beta + 1)]$,

$$\frac{(\beta + n + 1)(\beta + n)}{n(n+1)} - a\frac{\beta + n}{n} = -b\frac{\beta + n}{n} + c.$$

The solution of these equations is

$$b = \frac{(\alpha^2 - \beta^2)(\alpha + \beta + 2n + 1)}{2(n+1)(\alpha + \beta + 2n)(\alpha + \beta + n + 1)} \tag{11.8.3}$$

and

$$c = -\frac{(\alpha + n)(\beta + n)(\alpha + \beta + 2n + 2)}{(n+1)(\alpha + \beta + 2n)(\alpha + \beta + n + 1)}. \tag{11.8.4}$$

We show in the appendix that this result can also be obtained using the generating function (11.7.5).

11.9 Differential Relations

It was shown in chapter 1 that an expression for the derivative of $P_N^{(\alpha,\beta)}(x)$ can be written in the form

$$(1 - x^2)\frac{dP_N^{(\alpha,\beta)}(x)}{dx} = -NxP_N^{(\alpha,\beta)}(x) + b_N P_N^{(\alpha,\beta)}(x) + b_{N-1}P_{N-1}^{(\alpha,\beta)}(x). \tag{11.9.1}$$

Equating powers of x^N on both sides of the equation produces
$$b_N = \frac{N(\alpha - \beta)}{\alpha + \beta + 2N}. \tag{11.9.2}$$
If we put $x = 1$ and $x = -1$ in Eq. (11.9.1), we find
$$0 = (-N + b_N)\frac{\Gamma(N + \alpha + 1)}{\Gamma(\alpha + 1)N!} + b_{N-1}\frac{\Gamma(N + \alpha)}{\Gamma(\alpha + 1)(N - 1)!}$$
and
$$0 = (N + b_N)\frac{\Gamma(N + \beta + 1)}{\Gamma(\beta + 1)N!} - b_{N-1}\frac{\Gamma(N + \beta)}{\Gamma(\beta + 1)(N - 1)!}.$$
The solution of these two equations gives the previous value for b_N and
$$b_{N-1} = \frac{2(N + \alpha)(N + \beta)}{\alpha + \beta + 2N}. \tag{11.9.3}$$

This result can also be obtained using the generating function. We can show that
$$\left(\alpha + \beta + 2t\frac{\partial}{\partial t}\right)\left[(1 - x^2)\frac{\partial w}{\partial x} + xt\frac{\partial w}{\partial t}\right]$$
$$= (\alpha - \beta)t\frac{\partial w}{\partial t} + 2t^{1-\alpha}\frac{\partial}{\partial t}\left\{t^{\alpha-\beta+1}\frac{\partial(t^{\beta+1}w)}{\partial t}\right\}.$$
This is most easily done using a computer algebra package such as Maple. If we now substitute the expansion for $w(x, t)$ in (11.7.5), and equate powers of t^N on both sides, we obtain the differential relation
$$(\alpha + \beta + 2N)\left[(1 - x^2)\frac{d\,P_N^{(\alpha,\beta)}(x)}{dx} + NxP_N^{(\alpha,\beta)}(x)\right]$$
$$= (\alpha - \beta)NP_N^{(\alpha,\beta)}(x) + 2(N + \alpha)(N + \beta)P_{N-1}^{(\alpha,\beta)}(x). \tag{11.9.4}$$

11.10   Step Up and Step Down Operators

Rewriting Eq. (11.9.4) gives the step down operator S_N^-:
$$P_{N-1}^{(\alpha,\beta)}(x) = S_N^- P_N^{(\alpha,\beta)}(x)$$
$$= \left\{\frac{N(\beta - \alpha + (\alpha + \beta + 2N)x)}{2(N + \alpha)(N + \beta)} + \frac{(\alpha + \beta + 2N)(1 - x^2)}{2(N + \alpha)(N + \beta)}\frac{d}{dx}\right\}P_N^{(\alpha,\beta)}(x). \tag{11.10.1}$$
If we now combine this with the recurrence relation (11.8.1), we get the step up operator S_N^+:
$$P_{N+1}^{(\alpha,\beta)}(x) = S_N^+ P_N^{(\alpha,\beta)}(x)$$
$$= \left\{\frac{\alpha - \beta + (\alpha + \beta + 2N + 2)x}{2(N + 1)} - \frac{(\alpha + \beta + 2N + 2)(1 - x^2)}{2(N + 1)(\alpha + \beta + N + 1)}\frac{d}{dx}\right\}P_N^{(\alpha,\beta)}(x). \tag{11.10.2}$$

11.11   Appendix

Alternative derivation of the recurrence relation using the generating function:

Let

$$R = \sqrt{1 - 2xt + t^2}$$

and

$$w = \frac{1}{R(1 - t + R)^\alpha (1 + t + R)^\beta}.$$

Also let

$$r1 = 4t^{-(\alpha+\beta)} \frac{\partial}{\partial t} \left\{ t^{(\alpha+\beta)/2+2} \left[\frac{\partial}{\partial t} \left(t^{(\alpha+\beta)/2} \frac{\partial w}{\partial t} \right) \right] \right\}$$

$$= \sum_{n=1}^{\infty} 2n(\alpha + \beta + n)(\alpha + \beta + 2n - 2) P_n^{(\alpha,\beta)}(x) t^{n-1}.$$

$$r2 = (\alpha^2 - \beta^2) \left((\alpha + \beta + 1)w + 2t \frac{\partial w}{\partial t} \right)$$

$$= \sum_{n=0}^{\infty} (\alpha^2 - \beta^2)(\alpha + \beta + 2n + 1) P_n^{(\alpha,\beta)}(x) t^n.$$

$$r3 = 8xt^{(1-\alpha-\beta)/2} \frac{\partial}{\partial t} \left[t^{3/2} \frac{\partial^2}{\partial t^2} \left(t^{(\alpha+\beta)/2+1} w \right) \right]$$

$$= x \sum_{n=0}^{\infty} (\alpha + \beta + 2n)(\alpha + \beta + 2n + 1)(\alpha + \beta + 2n + 2) P_n^{(\alpha,\beta)}(x) t^n.$$

$$r4 := 2t^{-(\alpha+\beta)/2} \frac{\partial}{\partial t} \left\{ t^{(\alpha-\beta)/2+2} \frac{\partial}{\partial t} \left[t^{1+\beta-\alpha} \frac{\partial}{\partial t} \left(t^{\alpha+1} w \right) \right] \right\}$$

$$= 2 \sum_{n=0}^{\infty} (\alpha + n + 1)(\beta + n + 1)(\alpha + \beta + 2n + 4) P_n^{(\alpha,\beta)}(x) t^{n+1}.$$

It can be shown that $r_1 - r_2 - r_3 + r_4 = 0$. This is best done using a computer algebra package. If we then equate coefficients of t^n to zero, we obtain

$$2(n + 1)(\alpha + \beta + n + 1)(\alpha + \beta + 2n) P_{n+1}^{(\alpha,\beta)}(x)$$

$$= (\alpha + \beta + 2n + 1) \left[(\alpha^2 - \beta^2) + x(\alpha + \beta + 2n)(\alpha + \beta + 2n + 2) \right] P_n^{(\alpha,\beta)}(x))$$

$$-2(\alpha + n)(\beta + n)(\alpha + \beta + 2n + 2) P_{n-1}^{(\alpha,\beta)}(x),$$

which is the same as the relations (11.8.1), (11.8.2), (11.8.3) and (11.8.4) we obtained earlier.

References

Courant R and Hilbert D, Methods of Mathematical Physics Vol 1, Interscience Publishers, 1953.

Dennery P and Krzywicki A, Mathematics for Physicists, Harper and Rowe, 1967.

Erdelyi A, Higher Transcendental Functions Vol 2, McGraw-Hill, 1953.

Hochstrasser Urs W, Orthogonal polynomials, Chapter 22, Handbook of Mathematical Functions, Eds. Abramowitz M and Stegun I A, Dover, 1970.

Koornwinder T H, Wong R, Koekoek R and Swarttouw R F, Chapter 18, NIST Handbook of Mathematical Functions, Eds. Olver W J, Lozier D W, Boisvert R F and Clark C W, NIST and Cambridge University Press, 2009.

Chapter 12

General Appendix

In this appendix we detail the properties of a number of the functions used extensively in the previous chapters.

12.1 The Gamma Function

12.1.1 Definition

The Gamma function $\Gamma(z)$ is commonly defined for $\Re(z) > 0$ by the integral

$$\Gamma(z) = \int_0^\infty t^{z-1} e^{-t} dt. \tag{G1}$$

Integration by parts leads to the relation

$$\Gamma(z+1) = z\Gamma(z). \tag{G2}$$

From this relation we see that for n a positive integer, $\Gamma(n+1) = n!$.

The relation (G2) can be used to analytically continue the definition of $\Gamma(z)$ to all complex values of z except when z is a negative real integer or zero where it has a simple pole. The residue at $z = 0$ is 1 and that at $z = -n$, $(-1)^n/n!$.

An important special value is $\Gamma(1/2)$. Writing $t = s^2$, we see that

$$\Gamma(1/2) = \int_0^\infty t^{-1/2} e^{-t} dt = 2 \int_0^\infty e^{-s^2} ds$$

and

$$\left[\Gamma(1/2)\right]^2 = 4 \int_0^\infty e^{-x^2} dx \int_0^\infty e^{-y^2} dy.$$

If we now change to polar coordinates and combine the integrals,

$$\left[\Gamma(1/2)\right]^2 = 4 \int_0^\infty e^{-r^2} r\, dr \int_0^{\pi/2} d\theta = \pi.$$

Therefore $\Gamma(1/2) = \sqrt{\pi}$.

The change to polar coordinates can be justified by considering the limit as $R \to \infty$ of

$$4 \int_0^R e^{-x^2} dx \int_0^R e^{-y^2} dy.$$

Since e^{-r^2} is positive throughout the region of integration,

$$\int_0^R e^{-r^2} r \, dr \int_0^{\pi/2} d\theta < \int_0^R e^{-x^2} dx \int_0^R e^{-y^2} dy < \int_0^{\sqrt{2}R} e^{-r^2} r \, dr \int_0^{\pi/2} d\theta.$$

The difference between the third and first of the integrals above is

$$\frac{\pi}{4} \int_R^{\sqrt{2}R} e^{-r^2} r \, dr = \frac{\pi}{4}(e^{-R^2} - e^{-2R^2})$$

which tends to zero as $R \to \infty$.

12.1.2 Alternative Definition

An alternative definition of the Gamma function comes from Euler's formula (for example see Copson, p. 209). If we let

$$\Gamma(z, n) = \frac{n! \, n^z}{z(z+1)(z+2)...(z+n)} \qquad (n = 1, 2, 3...) \qquad \text{(G3)}$$

then we can show that in the limit $n \to \infty$, $\Gamma(z, n) \to \Gamma(z)$. Firstly we note that

$$\Gamma(z, n) = \left(\frac{n}{n+1)}\right)^z \prod_{r=1}^n \left\{ \left(1 + \frac{1}{r}\right)^z \left(1 + \frac{z}{r}\right)^{-1} \right\}.$$

The product inside the curly brackets

$$\left(1 + \frac{1}{r}\right)^z \left(1 + \frac{z}{r}\right)^{-1} = 1 + \frac{z(z-1)}{r^2} + O\left(\frac{1}{r^3}\right).$$

Noting that the product $\Pi_{r=1}^n (1 + a_r)$ converges absolutely if the series $\Sigma_{r=0}^n a_r$ converges absolutely, (see for example Titchmarsh, p. 15), we see that $\Gamma(z, n)$ tends to a finite limit as $n \to \infty$.

Integrating by parts the integral

$$\int_0^1 (1-s)^n s^{z-1} ds = \frac{n}{z} \int_0^1 s^z (1-s)^{n-1} ds$$

and so integrating this by parts a total of n times produces the function $\Gamma(z, n)/n^z$. If we write $s = u/n$, we get

$$\Gamma(z, n) = \int_0^n \left(1 - \frac{u}{n}\right)^n u^{z-1} du. \qquad \text{(G4)}$$

As $n \to \infty$, $(1 - u/n)^n \to \exp(-u)$ and so $\Gamma(z,n) \to \Gamma(z)$.

To prove this rigorously consider

$$f(u) = 1 - e^u \left(1 - \frac{u}{n}\right)^n.$$

For $0 \le u \le n$,

$$f'(u) = e^u \left(1 - \frac{u}{n}\right)^{n-1} \frac{u}{n} \ge 0.$$

Then

$$f(u) = \int_0^u f'(v)dv = \int_0^u e^v \left(1 - \frac{v}{n}\right)^{n-1} \frac{v}{n} dv \le e^u \int_0^u \left(1 - \frac{v}{n}\right)^{n-1} \frac{v}{n} dv$$

$$< \frac{e^u}{n} \int_0^u v\, dv = e^u \frac{u^2}{2n}$$

and therefore

$$0 \le e^{-u} - \left(1 - \frac{u}{n}\right)^n \le \frac{u^2}{2n}. \tag{G5}$$

Let us now split the integration range 0 to n into 0 to a and a to n and write

$$\int_0^n \left\{ e^{-u} - \left(1 - \frac{u}{n}\right)^n \right\} u^{z-1} du$$

$$= \int_0^a \left\{ e^{-u} - \left(1 - \frac{u}{n}\right)^n \right\} u^{z-1} du + \int_a^n e^{-u} u^{z-1} du - \int_a^n \left(1 - \frac{u}{n}\right)^n u^{z-1} du.$$

If we now choose a so large that for all values of n,

$$\left| \int_a^n e^{-u} u^{z-1} du \right| < \epsilon$$

and then also

$$\left| \int_a^n \left(1 - \frac{u}{n}\right)^n u^{z-1} du \right| < \left| \int_a^n e^{-u} u^{z-1} du \right| < \epsilon.$$

This follows from the inequality

$$\left(1 - \frac{u}{n}\right)^n < e^{-u}$$

which follows from inequality (G5) or can be proved by taking logarithms on both sides. Then

$$\left| \int_0^n \left\{ e^{-u} - \left(1 - \frac{u}{n}\right)^n \right\} u^{z-1} du \right| < \left| \int_0^a \left\{ e^{-u} - \left(1 - \frac{u}{n}\right)^n \right\} u^{z-1} du \right| + 2\epsilon$$

$$< \left| \frac{a^{z+2}}{2n(z+2)} \right| + 2\epsilon.$$

As this can be made arbitrarily small, this proves the limit $\lim_{n \to \infty} \Gamma(z,n) = \Gamma(z)$.

12.1.3   Duplication Formula

This result can be easily derived for integer values of the argument.

$$\Gamma(2n) = (2n-1)! = (2n-1)(2n-2)(2n-3)(2n-4)...2.1$$

$$= 2^{2n-1}(n-1/2)(n-1)(n-3/2)(n-2)...1.(1/2)$$

$$= 2^{2n-1}\Gamma(n+1/2)(n-1)!/\Gamma(1/2) = 2^{2n-1}\Gamma(n+1/2)\Gamma(n)/\sqrt{\pi}. \quad \text{(G6)}$$

For other values of the argument, we use the product formula for $\Gamma(z)$. We firstly note that

$$\Gamma(z,n) = \frac{n^z\Gamma(n+1)\Gamma(z)}{\Gamma(n+1+z)}$$

and hence that

$$\frac{n^z\Gamma(n)}{\Gamma(n+z)} = \frac{n+z}{n}\frac{\Gamma(z,n)}{\Gamma(z)} \to 1 \qquad \text{as} \qquad n \to \infty.$$

Consider now

$$\Gamma(2z,2n) = \frac{(2n)!(2n)^{2z}}{2z(2z+1)(2z+2)...(2z+2n)}$$

$$= \frac{2^{2z-1}n^{2z}n!\Gamma(n+1/2)}{\sqrt{\pi}z(z+1)(z+2)...(z+n)(z+1/2)(z+3/2)...(z+n-1/2)}$$

$$= \frac{2^{2z-1}}{\sqrt{\pi}}\Gamma(z,n)\Gamma(z+1/2,n)\left[\frac{\Gamma(n+1/2)}{n^{1/2}\Gamma(n)}\right]\left[\frac{(z+n+1/2)}{n}\right].$$

If we take the limit $n \to \infty$ we see

$$\Gamma(2z) = 2^{2z-1}\Gamma(z)\Gamma(z+1/2)/\sqrt{\pi}. \quad \text{(G7)}$$

12.2   The Beta Function

12.2.1   Definition

The Beta Function

$$B(a,b) = \int_0^1 t^{a-1}(1-t)^{b-1}dt \qquad \Re a,b > 0. \quad \text{(B1)}$$

Alternative representations for $\Re a,b > 0$ are

$$B(a,b) = \int_1^\infty s^{-a-b}(s-1)^{b-1}ds \quad \text{(B2)}$$

and

$$B(a,b) = 2\int_0^{\pi/2} \sin^{2a-1}\theta \cos^{2b-1}\theta d\theta \quad \text{(B3)}$$

obtained by substituting $t = 1/s$ and $t = \sin^2\theta$ respectively.

12.2.2   Expression in terms of Gamma Functions

Consider

$$\Gamma(a)\Gamma(b) = \int_0^\infty s^{a-1}e^{-s}ds \int_0^\infty t^{b-1}e^{-t}dt$$

$$= 4\int_0^\infty x^{2a-1}e^{-x^2}dx \int_0^\infty y^{2b-1}e^{-y^2}dy,$$

on replacing s by x^2 and t by y^2. If we now change to polar coordinates and write $x = r\cos\theta$ and $y = r\sin\theta$, we find

$$\Gamma(a)\Gamma(b) = 4\int_0^\infty r^{2a+2b-1}e^{-r^2}dr \int_0^{\pi/2} \cos^{2a-1}\theta \sin^{2b-1}\theta d\theta = \Gamma(a+b)B(a,b).$$

Therefore

$$B(a,b) = \frac{\Gamma(a)\Gamma(b)}{\Gamma(a+b)}. \tag{B4}$$

The justification of the change to polar coordinates and changes in the integration limits for real positive a and b is essentially the same as that in the case of the calculation of $\Gamma(1/2)$. Specifically, writing $f(x,y) = x^{2a-1}y^{2b-1}$ and noting that $e^{-r^2}f(x,y)$ is positive throughout the integration region

$$\int_0^R e^{-r^2}rdr \int_0^{\pi/2} f(x,y)d\theta < \int_0^R e^{-x^2}dx \int_0^R e^{-y^2}dy f(x,y)$$

$$< \int_0^{\sqrt{2}R} e^{-r^2}rdr \int_0^{\pi/2} f(x,y)d\theta.$$

In the limit $R \to \infty$, the difference between the third and first of the integrals above tends to zero confirming the convergence.

If $a = 1 - b$, $\Gamma(a+b) = 1$ and so

$$\Gamma(a)\Gamma(1-a) = B(a, 1-a) = \int_0^1 t^{a-1}(1-t)^{-a}dt.$$

We are supposing for the moment that $0 < \Re a < 1$.

Writing $s = t/(1-t)$ gives

$$\Gamma(a)\Gamma(1-a) = \int_0^\infty \frac{s^{a-1}}{1+s}ds.$$

This can be evaluated using a contour integral. Consider

$$\int_c \frac{s^{a-1}}{1+s}ds,$$

where the contour consists of the integral from ϵ to R just above the real axis, anticlockwise along the circle of radius R to just below the real axis, the integral from R to ϵ just below the real axis and the circle of radius ϵ clockwise to the starting point. There is one pole inside the contour at $s = -1$. Along the first part of the contour, we have

$$\int_\epsilon^\infty \frac{s^{a-1}}{1+s} ds.$$

Along a circle of radius ρ, we have with $s = \rho e^{i\theta}$,

$$\int_0^{2\pi} \frac{\rho^a e^{ia\theta}}{\rho e^{i\theta} + 1} i d\theta.$$

Along the circle of radius R, since $\Re a < 1$, this integral tends to zero as $R \to \infty$ and along the circle of radius ϵ, the integral tends to zero as $\epsilon \to 0$ since $\Re a > 0$. The other integral along the real axis becomes

$$-e^{2\pi i(a-1)} \int_\epsilon^R \frac{s^{a-1}}{1+s} ds.$$

Thus

$$(1 - e^{2\pi ai}) \int_0^\infty \frac{s^{a-1}}{1+s} ds = 2\pi i e^{\pi i(a-1)}.$$

Therefore

$$\Gamma(a)\Gamma(1-a) = B(a, 1-a) = \int_0^\infty \frac{s^{a-1}}{1+s)} ds = \frac{\pi}{\sin \pi a}. \tag{B5}$$

This relation can be extended by noting that

$$\Gamma(1+a)\Gamma(-a) = a\Gamma(a)\Gamma(1-a)/a.$$

This means that we can remove the restriction $0 < \Re a < 1$ and replace it with $a \neq$ an integer.

12.2.3 Contour Integral Representations of the Beta Function

Consider the contour integral

$$\int_{c1} t^{a-1}(1-t)^{b-1} dt \qquad 0 < \Re a, \tag{B6}$$

where the contour $c1$ is a loop starting at $t = 0$ on the real axis, passing round the point $t = 1$ in an anticlockwise direction and ending up back at the origin.

Contour $c1$

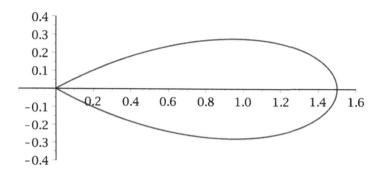

This contour can be shrunk onto an integral along the real axis from $t = 0$ to $1 - \epsilon$, a circle of radius ϵ anticlockwise about the point $t = 1$ and an integral from $1 - \epsilon$ back to 0. We have

$$\int_0^{1-\epsilon} t^{a-1}(1-t)^{b-1}dt - i\epsilon^b \int_0^{2\pi} (1-\epsilon\, e^{i\theta})^{a-1} e^{ib\theta} d\theta + e^{2\pi ib} \int_{1-\epsilon}^0 t^{a-1}(1-t)^{b-1}dt.$$

For $\Re b > 0$, we can shrink the integral of radius ϵ about $t = 1$. This limit tends to zero as $\epsilon \to 0$. This leaves

$$\int_{c1} t^{a-1}(1 - t)^{b-1}dt = (1 - e^{2\pi ib}) \int_0^1 t^{a-1}(1 - t)^{b-1}dt = (1 - e^{2\pi ib})B(a,b).$$

Therefore

$$B(a, b) = i\frac{e^{-\pi ib}}{2\sin \pi b} \int_{c1} t^{a-1}(1 - t)^{b-1}dt \qquad \Re a > 0. \qquad (B7)$$

The integral in Eq. (B7) is defined for all values of b except when b is an integer and thus represents the analytic continuation of the Beta Function.

Another contour integral representation which can be analytically continued to negative values of a is the contour integral

$$\int_{c2} t^{a-1}(1 - t)^{b-1}dt \qquad 0 < \Re b, \qquad (B8)$$

where the contour $c2$ is a loop starting at $t = 1$ on the real axis, passing round the origin in an anticlockwise direction and ending up at $t = 1$.

Contour $c2$

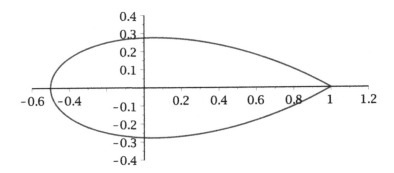

This contour can be shrunk onto an integral just above the real axis from 1 to ϵ, a circle of radius ϵ anticlockwise about the origin and an integral from ϵ to 1 just below the real axis. We have

$$\int_1^\epsilon t^{a-1}(1-t)^{b-1}dt + i\epsilon^a \int_0^{2\pi} e^{ia\theta}(1-\epsilon e^{i\theta})^{b-1}d\theta + e^{2\pi ia}\int_\epsilon^1 t^{a-1}(1-t)^{b-1}dt.$$

If $0 < \Re\,a$, the integral round the circle of radius ϵ tends to zero as $\epsilon \to 0$. This leaves

$$(e^{2\pi ia} - 1)\int_0^1 t^{a-1}(1-t)^{b-1} = (e^{2\pi ia} - 1)B(a,b).$$

Therefore

$$B(a,b) = \frac{e^{-\pi ia}}{2i\sin(\pi a)}\int_{c2} t^{a-1}(1-t)^{b-1}dt. \qquad \Re\,b > 0. \qquad (B9)$$

The integral on the right of Eq. (B9) is defined for negative $\Re\,a$ and so represents the analytic continuation of the Beta Function for non-integer values of a.

12.3   The Hypergeometric Function

12.3.1   *Definition*

The Hypergeometric Function $_2F_1(a,b;c;z)$ is defined for $|z| < 1$ by the series

$$_2F_1(a,b;c,z) = 1 + \frac{ab}{c}z + \frac{a_2b_2)}{c_2}\frac{z^2}{2!} + \frac{a_3b_3}{c_3}\frac{z^3}{3!} + \dots$$

$$= \frac{\Gamma(c)}{\Gamma(a)\Gamma(b)} \sum_{n=0}^{\infty} \frac{\Gamma(a+n)\Gamma(b+n))}{\Gamma(c+n)} \frac{z^n}{n!}, \tag{H1}$$

where we have used the Pochammer symbol, $a_n = a(a+1)\dots(a+n-1)$. This series converges absolutely if $|z| < 1$ as can be seen using the ratio test and also if $|z| = 1$ if $\Re(c-a-b) > 0$. This follows since the ratio of successive terms of the series for $|z| = 1$ is for large n

$$\left|\frac{u_{n-1}}{u_n}\right| \sim 1 + \frac{\Re(c-a-b)}{n} + O(1/n^2).$$

12.3.2   Integral Formulae

Equation (H1) can be written in the form

$$_2F_1(a,b;c;z) = \frac{\Gamma(c)}{\Gamma(a)\Gamma(b)\Gamma(c-b)} \sum_{n=0}^{\infty} \Gamma(a+n)B(b+n,c-b)\frac{z^n}{n!}. \tag{H2}$$

If we now replace the Beta function by its integral representation with $\Re c > \Re b > 0$,

$$_2F_1(a,b;c;z) = \frac{\Gamma(c)}{\Gamma(a)\Gamma(b)\Gamma(c-b)} \sum_{n=0}^{\infty} \int_0^1 t^{b+n-1}(1-t)^{c-b-1}\Gamma(a+n)\frac{z^n}{n!}.$$

We now interchange the summation and integration and note that

$$\sum_{n=0}^{\infty} \frac{\Gamma(a+n)}{\Gamma(a)} \frac{(zt)^n}{n!} = \frac{1}{(1-zt)^a}$$

so that

$$_2F_1(a,b;c;z) = \frac{\Gamma(c)}{\Gamma(b)\Gamma(c-b)} \int_0^1 \frac{t^{b-1}(1-t)^{c-b-1}}{(1-zt)^a} dt \qquad \Re c > b > 0. \tag{H3}$$

We obtain the value when $z = 1$ by taking the limit $z \to 1-$. This gives

$$_2F_1(a,b;c;1) = \frac{\Gamma(c)}{\Gamma(b)\Gamma(b-c)} \int_0^1 t^{b-1}(1-t)^{c-a-b-1} dt = \frac{\Gamma(c)\Gamma(c-a-b)}{\Gamma(c-a)\Gamma(c-b)} \tag{H4}$$

provided that $\Re b > 0$ and $\Re(c-a-b) > 0$.

For $|z| < 1$ there is no problem in extending the definition of the Hypergeometric Function $_2F_1(a,b;c;z)$ to negative values of a by Eq. (H3).

However in the calculations of the formal expressons of the expansion co-efficients of the Chebyshev polynomials of the third and fourth kinds, we come across Hypergeometric functions with negative values of b and a. In order to proceed we need to use contour integral representations of the Hypergeometric functions derived from those for the Beta function in Eq. (B6) and Eq. (B8). Proceeding as in Eq. (H2) with the contour integral representations for the Beta functions, we have

$$_2F_1(a, b; c; z) = \frac{\Gamma(c)}{\Gamma(b)\Gamma(c-b)} i \frac{e^{-\pi i(b-c)}}{2\sin\pi(b-c)} \int_{c1} \frac{t^{b-1}(1-t)^{c-b-1}}{(1-zt)^a} dt \quad (H5)$$

where $\Re\, b > 0$, $c - b \neq$ integer and $c1$ is the contour for Eq. (B6), and

$$_2F_1(a, b; c; z) = \frac{\Gamma(c)}{\Gamma(b)\Gamma(c-b)} \frac{e^{-\pi i b}}{2i\sin\pi b} \int_{c2} \frac{t^{b-1}(1-t)^{c-b-1}}{(1-zt)^a} dt, \quad (H6)$$

where $\Re(c - b) > 0$, $b \neq$ integer and $c2$ is the contour for Eq. (B8).

Using the reflection property of the Gamma function Eq. (B5) in the form

$$\Gamma(b)\sin\pi b = \frac{\pi}{\Gamma(1-b)},$$

we see that

$$_2F_1(a, b; c; z) = \frac{\Gamma(c)\Gamma(1-b)}{\Gamma(c-b)} \frac{e^{-\pi i b}}{2\pi i} \int_{c2} \frac{t^{b-1}(1-t)^{c-b-1}}{(1-zt)^a} dt. \quad (H7)$$

In this form the formula can be analytically continued to negative integer values of b. For non-positive integer values of $b = -m$ say, the loop integral in Eq. (H7) can be shrunk to a circle round the origin so that

$$_2F_1(a, -m; c; z) = \frac{(-1)^m \Gamma(c)}{\Gamma(c+m)} \frac{d^m}{dt^m} \left(\frac{(1-t)^{c+m-1}}{(1-zt)^a} \right)\Big|_{t=0}. \quad (H8)$$

For $z = 1$ this becomes

$$_2F_1(a, -m; c; 1) = \frac{(-1)^m \Gamma(c)}{\Gamma(c+m)} \frac{d^m}{dx^m} \left[(1-t)^{c+m-a-1} \right]$$

$$= \frac{\Gamma(c)\Gamma(c+m-a)}{\Gamma(c+m)\Gamma(c-a)} = \frac{(c-a)_m}{(c)_m}, \quad (H9)$$

a result commonly known as Vandermonde'e theorem. This extends the result of Eq. (H4) to negative integer values of $b = -m$.

Following Carlson, an alternative derivation of Vandermonde's theorem comes from considering the binomial expansion of

$$\frac{1}{(1-x)^{a+b}} = \sum_{k=0}^{\infty} \frac{(a+b)_k}{k!} x^k = \frac{1}{(1-x)^a} \frac{1}{(1-x)^b} = \sum_{i=0}^{\infty} \frac{a_i}{i!} x^i \sum_{j=0}^{\infty} \frac{b_j}{j!} x^j.$$

Equating the coefficients of x^n on both sides leads to

$$\frac{(a+b)_n}{n!} = \sum_{i=0}^{n} \frac{a_i\, b_{n-i}}{i!\,(n-i)!} = S$$

$$= \frac{b_n}{n!}\left\{1 + \frac{na}{(b+n-1)} + \frac{n(n-1)a(a+1)}{(b+n-1)(b+n-2)2!} + \cdots\right\},$$

and so the sum S in the curly brackets is

$$S = {}_2F_1(a,-n;1-n-b;1) = \frac{(a+b)_n}{b_n} = \frac{\Gamma(a+b+n)\Gamma(b)}{\Gamma(a+b)\Gamma(b+n)}. \qquad \text{(H10)}$$

If we multiply $(a+b)_n$ and b_n by $(-1)^n$,

$$\frac{(a+b)_n}{b_n} = \frac{(1-n-a-b)_n}{(1-n-b)_n} = \frac{\Gamma(1-b-a)\Gamma(1-n-b)}{\Gamma(1-b)\Gamma(1-n-b-a)}$$

which is of the same form as Eq. (H9).

There are a number of relations expressing a Hypergeometric Function for a particular value of z in terms of a Hypergeometric Function of a different argument. (See Oberhettinger or Olde Daalhuis.) We shall need the one relating z with $z/(1-z)$. If we write $s = 1 - t$ in the integrand of (H3), (H5) or (H6), the denominator becomes $(1-z)^a[1 - z/(z-1)]^a$. For Eq. (H3) we have

$$\frac{\Gamma(c)}{\Gamma(b)\Gamma(c-b)}(1-z)^{-a}\int_0^1 \frac{s^{c-b-1}(1-s)^{b-1}}{[1-tz/(z-1)]^a}\,ds.$$

Therefore

$${}_2F_1(a,b;c;z) = (1-z)^{-a}\,{}_2F_1(a,c-b;c;z/(1-z)). \qquad \text{(H11)}$$

For the integral in Eq. (H5), the substitution $s = 1 - t$ changes the contour $c1$ to $c2$ and for the integral in Eq. (H6), the contour $c2$ is changed into $c1$. We use the symmetry property of the hypergeometric function to write the right-hand side of Eq. (H10)

$$(1-z)^{-a}\,{}_2F_1\left(c-b,a;c;\frac{z}{z-1}\right) = (1-z)^{c-b-a}\,{}_2F_1(c-b;c-a;c;z).$$

Then we have

$${}_2F_1(a,b;c;z) = (1-z)^{c-a-b}\,{}_2F_1(c-a,c-b;c;z). \qquad \text{(H12)}$$

This relation can be used to prove Saalschutz's theorem. If we write the expansions of both sides in powers of z and equate the coefficients of z^n on both sides we find

$$\frac{a_n\, b_n}{c_n\, n!} = \sum_{j=0}^{n} \frac{(a+b-c)_{n-j}(c-a)_j(c-b)_j}{(n-j)!\, c_j\, j!}$$

$$= \frac{(a+b-c)_n}{n!} \left\{ 1 + \frac{(c-a)(c-b)n}{c(a+b-c+n-1)} + \cdots \right\}.$$

The sum in the curly brackets is the hypergeometric function

$$_3F_2(c-a, c-b, -n; c, c+1-a-b-n; 1) = \frac{a_n b_n}{c_n(a+b-c)_n}. \qquad \text{(H13)}$$

The Confluent Hypergeometric Function $_1F_1(\alpha; \beta; z)$ can be obtained from the Hypergeometric Function $_2F_1(a, b; c; z)$ by taking the limit $a \to \infty$

$$_1F_1(\alpha; \beta; z) = \lim_{a \to \infty} [_2F_1(a, \alpha; \beta; z/a)]. \qquad \text{(H14)}$$

This leads to further representations for the Laguerre polynomial $L_n^\alpha(x)$ by taking this limit in (H3), (H5), (H6) and (H8). In particular

$$L_n^{(\alpha)}(z) = \frac{(\alpha+1)_n}{n!} \, _1F_1(-n; \alpha+1; z) = \frac{(-1)^n}{n!} \frac{d^n}{dt^n} \left\{ (1-t)^{\alpha+n} e^{tz} \right\} \Big|_{t \to 0}.$$
$$\text{(H15)}$$

References

Askey R A, Roy R, Chapter 5, NIST Handbook of Mathematical Functions, Eds. Olver W J, Lozier D W, Boisvert R F and Clark C W, NIST and Cambridge University Press, 2009.

Carlson B C, Special Functions of Applied Mathematics, Academic Press, 1977.

Davis P J, Chapter 6, Handbook of Mathematical Functions, Eds. Abramowitz M and Stegun I A, Dover, 1967.

Copson E T, Theory of Functions of a Complex Variable, Oxford University Press, 1955.

Koornwinder T H, Wong R, Koekoek R and Swarttouw R F, Chapter 18, NIST Handbook of Mathematical Functions, Eds. Olver W J, Lozier D W, Boisvert R F and Clark C W, NIST and Cambridge University Press, 2009.

Macrobert T M, Functions of a Complex Variable, Macmillan, 1933.

Oberhettinger F, Chapter 15, Handbook of Mathematical Functions, Eds. Abramowitz M and Stegun I A, Dover, 1967.

Olde Daalhuis A B, Chapter 15, NIST Handbook of Mathematical Functions, Eds. Olver W J, Lozier D W, Boisvert R F and Clark C W, NIST and Cambridge University Press, 2009.

Sneddon I N, Special Functions of Mathematical Physics and Chemistry, Oliver and Boyd, 1956.

Titchmarsh E C, The Theory of Functions, Oxford University Press, 1958

Whittaker E T and Watson G N, A Course of Modern Analysis, Cambridge University Press, 1963.

Index